Praise for *SysML Distilled*

"In keeping with the outstanding tradition of Addison-Wesley's technical publications, Lenny Delligatti's *SysML Distilled* does not disappoint. Lenny has done a masterful job of capturing the spirit of OMG SysML as a practical, standards-based modeling language to help systems engineers address growing system complexity. This book is loaded with matter-of-fact insights, starting with basic MBSE concepts to distinguishing the subtle differences between use cases and scenarios to illumination on namespaces and SysML packages, and even speaks to some of the more esoteric SysML semantics such as token flows."

— *Jeff Estefan, Principal Engineer, NASA's Jet Propulsion Laboratory*

"The power of a modeling language, such as SysML, is that it facilitates communication not only within systems engineering but across disciplines and across the development life cycle. Many languages have the potential to increase communication, but without an effective guide, they can fall short of that objective. In *SysML Distilled,* Lenny Delligatti combines just the right amount of technology with a common-sense approach to utilizing SysML toward achieving that communication. Having worked in systems and software engineering across many domains for the last 30 years, and having taught computer languages, UML, and SysML to many organizations and within the college setting, I find Lenny's book an invaluable resource. He presents the concepts clearly and provides useful and pragmatic examples to get you off the ground quickly and enables you to be an effective modeler."

— *Thomas W. Fargnoli, Lead Member of the Engineering Staff, Lockheed Martin*

"This book provides an excellent introduction to SysML. Lenny Delligatti's explanations are concise and easy to understand; the examples well thought out and interesting."

— *Susanne Sherba, Senior Lecturer, Department of Computer Science, University of Denver*

"Lenny hits the thin line between a reference book for SysML to look up elements and an entertaining book that could be read in its entirety to learn the language. A great book in the tradition of the famous *UML Distilled.*"

— *Tim Weilkiens, CEO, oose*

"More informative than a PowerPoint, less pedantic than an OMG Profile Specification, *SysML Distilled* offers practicing systems engineers just the right level of the motivation, concepts, and notation of pure OMG SysML for them to attain fluency with this graphical language for the specification and analysis of their practical and complex systems."

— Lonnie VanZandt, chief architect, No Magic, Inc.

"Delligatti's *SysML Distilled* is a most aptly named book; it represents the distillation of years of experience in teaching and using SysML in industrial settings. The author presents a very clear and highly readable view of this powerful but complex modeling language, illustrating its use via easy-to-follow practical examples. Although intended primarily as an introduction to SysML, I have no doubt that it will also serve as a handy reference for experienced practitioners."

— Bran Selic, president, Malina Software Corp.

"SysML is a rather intimidating modeling language, but in this book Lenny makes it really easy to understand, and the advice throughout the book will help practitioners avoid numerous pitfalls and help them grasp and apply the core elements and the spirit of SysML. If you are planning on applying SysML, this is the book for you!"

— Celso Gonzalez, senior developer, IBM Rational

"*SysML Distilled* is a great book for engineers who are starting to delve into model-based systems engineering. The space system examples capture the imagination and express the concepts in a simple but effective way."

— Matthew C. Hause, chief consulting engineer, Atego and chair, OMG UPDM Group

"I've been deeply involved with OMG since the 1990s, but my professional needs have not often taken me into the SysML realm. So I thought I'd be a good beta tester for Lenny's book. To my delight, I learned a great deal reading through it, and I know you will too."

— Doug Tolbert, distinguished engineer, Unisys, and member, OMG Board of Directors and Architecture Board

"*SysML Distilled* provides a clear and comprehensive description of the language component of model-based systems engineering, while offering suggestions for where to find information about the tool and methodology components. There is evidence throughout the book that the author has a deep understanding of SysML and its application in a system development process. I will definitely be using this as a textbook in the MBSE courses I teach."

— J. D. Baker, OCUP, OCSMP, member of the OMG Architecture Board

"*SysML Distilled* is the desktop companion that many SysML modelers have needed for their bookshelves. Lenny has the experience and certifications to help you through your day-to-day modeling questions. This book is not a tutorial, nor is it the encyclopedic compendium of all things SysML. If you model using SysML, this will become your daily companion, as it is meant to be used regularly. I believe your copy will soon be dog-eared, with sticky notes throughout."

— Dr. Robert Cloutier, Stevens Institute of Technology

"SysML is utilized today in a wide range of applications, including deep space robotic spacecraft and down-to-earth agricultural equipment. This book concisely presents SysML in a manner that is both refreshingly accessible for new learners and quite handy for seasoned practitioners."

— Russell Peak, MBSE branch chief, Aerospace Systems Design Lab, Georgia Tech

"*SysML Distilled* is a wonderfully written, knowledgeable, and concise addition to systems modeling literature. The lucid explanations lead a newcomer by the hand into modeling reasonably complex systems, and the wealth and depth of the coverage of the most-used aspects of the SysML modeling language stretch to even enabling advanced intermediate depictions of most systems. It also serves as a handy reference. Kudos to Mr. Delligatti for gifting the world with this very approachable view of systems modeling."

— Bobbin Teegarden, CTO/chief architect, OntoAge and Board Member, No Magic, Inc.

SysML Distilled

SysML Distilled

A Brief Guide to the Systems Modeling Language

Lenny Delligatti

Addison-Wesley

Upper Saddle River, NJ • Boston • Indianapolis • San Francisco
New York • Toronto • Montreal • London • Munich • Paris • Madrid
Capetown • Sydney • Tokyo • Singapore • Mexico City

Library of Congress Cataloging-in-Publication Data
Delligatti, Lenny.
SysML distilled : a brief guide to the systems modeling language / Lenny Delligatti.
pages cm
Includes bibliographical references and index.
ISBN-13: 978-0-321-92786-6 (paperback : alk. paper)
ISBN-10: 0-321-92786-9 (paperback : alk. paper)
1. Systems engineering—Data processing. 2. Engineering systems—Computer simulation. 3. SysML (Computer science) I. Title.
TA168.D44 2014
620.00285'5133—dc23 2013035922

ISBN-13: 978-0-321-92786-6
ISBN-10: 0-321-92786-9

Text printed in the United States on recycled paper at LSC Communications in Crawfordsville, Indiana.

13 2020

This book is dedicated to my wife, Natalie, and my children, Aidan and Noelle—my greatest blessings . . . and my reasons for enduring the many late nights of writing.

Contents

Foreword by Rick Steiner

Systems engineering is not an easy subject to teach. Earlier in my career, I was emphatically told that systems engineering could not be taught in a classroom and that it could only be learned through experience. While that hasn't proven to be true, there are certainly concepts within the practice of systems engineering that are both subtle and arcane.

Expressing these concepts in models demands a suitably robust language, which is why a dedicated group of us began development of what would become SysML in early 2002. We attempted to be parsimonious and direct when designing the language, specifically targeting it for use by practicing systems engineers. I'm convinced that the resulting language is both flexible and useful, and I am gratified that it has emerged as a dominant standard for communicating systems-related ideas.

Just like the practice of systems engineering, however, SysML has proven difficult to teach effectively. The scope of systems engineering is remarkably broad, and even though SysML is a relatively compact language, students frequently get overwhelmed with its complexity. Resources for learning SysML and model-based systems engineering have until recently been rather limited, but it's getting better. Formal MBSE and SysML courses are now regularly being taught through several university or extension catalogs, and at least one comprehensive textbook is now available.

An engineer or manager who wants to casually learn the basics of SysML isn't likely to want to take a class. An advanced systems engineer who finds him- or herself in the middle of a project with tight deadlines just doesn't have the time to take a class. It is in both of these situations that this book has the greatest value.

Structured in a manner similar to Martin Fowler's popular *UML Distilled,* this book lays out the fundamentals of SysML diagrams in clear, concise terms. It is written in a casual, lighthearted manner, yet it conveys the gist of each concept and its graphical representation. What I like best about this book is that it keeps me reading, without getting bogged down in "meta-speak" and "UML-isms." It is sprinkled with humor and practical advice.

This is not a textbook or guidebook for SysML application or MBSE deployment, nor does it describe in detail the methodological rationale for each of the systems engineering concepts it describes. While it does use a consistent satellite example through the chapters, it does not walk the reader through any particular MBSE process. It is not a workbook, nor does it include questions or sample problems for the reader to work out. You as a SysML user or advanced MBSE practitioner may eventually need these other resources, but this book is an excellent start.

This book is a solid, self-paced, lightweight SysML reference guide. The world is ready for this book.

—*Rick Steiner,*
coauthor, A Practical Guide to SysML

Foreword by Richard Soley

Technology Take-Up Takes Time

I had the great luck to attend one of the best technologically focused (and entrepreneurially focused) universities in the world in the 1970s and 1980s. The future, as Steve Jobs might have put it, was invented there, not discovered. It was one of the places where "hackers stayed up late" and helped to create radar, flash photography, and the Internet. Those technologies helped change the world; more importantly, economies flourished through the creation of companies and other organizations that put those technologies to work. The computing explosion that started in the 1960s certainly fared well in the Massachusetts of 1980.

My own contributions during my initial foray into the academic world, eleven years at the Massachusetts Institute of Technology, moved and changed as my academic interests moved and changed, starting with the artificial intelligence field (handwriting recognition was an early focus), moving on in graduate school to computing systems architectures, and finally melding those two interests. Not a small contribution to my focus was being involved in five start-ups during my MIT years (though perhaps it was a large contribution to the length of time I spent at MIT). Artificial intelligence pioneers like Symbolics and Gold Hill Computer were important to my understanding of the *application* of technology; and my own start-up, A.I. Architects (with likely the best systems engineer I have ever met), strongly depended on the collision between the demands of artificial intelligence and the limited computing power of the early personal computing revolution.

Probably the most important single idea that I learned during this period was that the time it takes for technology to come out of the laboratory and into production is far greater than any academic believes. The expert systems of the 1980s, now a primary fixture of diagnostic and other systems worldwide (though generally under the moniker of "rule-based systems"), were clearly based on systems like PLANNER and CONNIVER from the 1960s. Twenty years seemed like the right rule of thumb; taking a technology through the engineering requirements necessary to stabilize and replicate the approach on an industrial scale, to the market development and integration, takes time.

OMG Objects of Awe

Nevertheless, when the Object Management Group (OMG) started up in 1989, the promise of *object technology* and *distributed objects* was to change the face of computing. As the Internet slowly changed into the World Wide Web, it was clear that consistent, standardized middleware would make it more possible than ever to integrate not only text pages from around the world but also application interoperability. The ability to "mash up" (as we would say twenty years later) computing power and data sources worldwide, using standardized APIs and on-the-wire protocols, would be far simpler with an object-oriented approach.

While OMG did a good job from the beginning in controlling the hype, avoiding the "artificial intelligence winter" that arose from an overhyped AI market in the 1980s, OMG likely didn't do a good enough job of recognizing that technology take-up takes time. It would be fifteen or twenty years before mash-ups became mash-ups, and object-oriented languages (initially C++, itself a good twenty years after Simular; now Java, C#, and Ruby-on-Rails) would permeate the computing milieu. OMG's objects of awe, as with all technologies, would become the quotidian tools of software developers everywhere, but it would take a couple of decades.

Modeling Makes Mavens

In the meantime, another opportunity would come OMG's way, with the proposal in 1996 that the object-oriented analysis and design market (as it was then called) had reached a dead end, an impasse, based not on the inherent technology but rather on the multitudinous approaches (and worse, notations) flooding the market. Even those technology mavens in love with the approach found themselves stymied by too much choice (and too little guarantee of portability and interoperability, once a choice was made). The mid-1990s consolidation of the analysis and design market created a vendor-focused market force for the creation of a standard, a force that was widely accepted by the slow-growing user community. The creation of the Unified Modeling Language (UML) standard in 1997, even with only a shared notation and not a shared methodology, was sufficient to coax the market into more than 100 percent CAGR over the next decade and a half. As the application development life cycle is more than just analysis and design, including also development, test, implementation, and maintenance, what had been just for "analysis and design" was soon called *modeling*.

And proof points for modeling, even with nascent standards like UML, abounded within a few years. Early scientific analysis showed 35 percent or more productivity increases using a modeling approach (as opposed to low-level programming language development); perhaps more importantly, as maintenance and support range from 80 percent to 90 percent of the software development life cycle, a couple of key analyses showed that 35 percent productivity increases (or better) could be had *in maintenance and integration*. This acceptance—as of this writing, according to market analysts Gartner & Forrester, including more than 71 percent of all software development teams—led to an explosion of modeling-related standardization at OMG.

Within fifteen years of the availability of the OMG UML standard (and its associated and very powerful parent, the Meta Object Facility, MOF), a fleet of domain-specific modeling languages were standardized. Languages and profiles for defining systems on a chip, for service-oriented architectures (SoaML), for business modeling and analysis (BPMN), for capturing enterprise architectures (UPDM), for defining rule-based systems (SBVR), even for capturing the motivations behind systems development (BMM), all joined the OMG stable. More importantly, most work at OMG shifted to "vertical markets," addressing the needs of healthcare information technology, financial services, life sciences, automotive and other consumer device dependability analysis, and so forth—all based on a view of systems based on high-level models.

Servicing the Spread of Systems

One of the most important horses in that stable is the OMG Systems Modeling Language, SysML. Defined as a "profile" of the UML, SysML took on the huge task of being the language that could integrate many disparate views of large-systems engineering: not only software and hardware but also requirements, mathematical parameterization, facilities management, design for maintenance, even the management of human and other resources and the behavior of the system under design. The vision I had outlined in 2001, called *model driven architecture*, could come to fruition with such an approach to integrated engineering, and not just for "software architecture," but for the overall structure of complex systems like aircraft carriers and chemical plants. As the IDEF series of specifications had promised in the early 1980s, SysML could truly bring together the expertise necessary from many fields to build well-designed, fit-to-purpose, and *maintainable* large systems.

So here we are, a dozen years removed from the first mention of model driven architecture and coming up on the requisite twenty years since the delivery of the Unified Modeling Language, with a book in hand that integrates the views of experts on how to think about and how to *use* SysML to deliver real systems. Here we find SysML *distilled*: according to the dictionary, metaphorically, its essential meaning or most important aspects extracted and displayed for all to see.

Complex systems development is, by its nature, a team sport. No one person can manage even the gathering of requirements for large systems; the size alone makes such a project complex. Since the real focus of design is simplification along one or more dimensions, we need notations and processes that not only communicate the simplified vision but also allow designers, developers, and engineers to drill down into a system's design and explore, in fractal fashion, the underlying parts of the design, the expectations and requirements, and the integration methodology. It's one thing to know that a notation like SysML—large and complex itself, of course, and including many different tools in its toolbox—can support large systems development; it's quite another to get past the learning curve to be able to effectively use those tools. My father-in-law was well known for using a screwdriver for every handyman task around the house (including driving nails); I prefer to have tools that are fit for purpose and to understand how to use those tools in an integrated way. Further, the SysML modeling language is not intended only to implement large complex systems but also to *communicate their design* to users of those systems; to maintainers of those systems; and to those who may have to debug and integrate extensions, corrections, and changes to those systems.

This book presents that introduction to the toolbox; better, it explains how to use those tools together to gather requirements for, build designs for, analyze designs of, and *communicate that process* to others in a design team (or future integration team). That's what engineers do, and SysML is the best way to do it.

—Richard Mark Soley, Ph.D.,
chairman and chief executive officer,
Object Management Group, Inc.

Preface

Why *SysML Distilled*? It's simple: You're busy. You need to know SysML now. You already have some systems modeling work to do. You don't need to know every detail of the language. You just want a book that focuses you on those parts of SysML that are most common and most useful in daily practice. *SysML Distilled* is that book.

You may choose to use this book as a desk reference, reaching for it when you're stuck and you've got a deadline bearing down on you. Or you may choose to dive deep one chapter at a time, adding new modeling skills to your toolbox for the future work coming your way. Or you may decide to read this book cover to cover to prepare for the first two levels of the *OMG Certified Systems Modeling Professional (OCSMP)* certification: *OCSMP Model User* and *OCSMP Model Builder: Fundamental*. This book is designed to serve you in all these ways.

Who Should Read This Book?

SysML is a graphical modeling language that you can use to visualize and communicate the designs of sociotechnical systems on all scales—systems consisting of hardware, software, data, people, and processes. Systems engineers are the ones who are responsible for the specification, analysis, design, verification, and validation of sociotechnical systems. Systems engineers—and students of systems engineering—are therefore the target audience for this book.

But that's an oversimplification. Many authors and teachers have repeated the axiom, "Everything is a system." Allow me to add the corollary: "Every engineer is a systems engineer." No matter your domain or job title, you've likely performed some or all of the systems engineering tasks I've mentioned. The premise of this book is that you can perform those activities more effectively via the standardized medium of an integrated SysML model than you can with nonstandardized modes of communication in disjoint sets of documents and diagrams. You are a systems engineer—and you want to do your job more effectively. *You* are therefore the target audience for this book.

What do you need to know before you dive in? You should have at least a conceptual understanding of system specification, analysis, design, verification, and validation. Knowing in advance what these activities consist of will help you internalize the ways SysML can help you do them better. The International Council on Systems Engineering (INCOSE) *Systems Engineering Handbook* is the authoritative reference.

You do not need to have any experience with any modeling language to benefit from this book. You may already know that SysML is based on the Unified Modeling Language (UML). In fact, you may have read Martin Fowler's seminal book, *UML Distilled*. I designed *SysML Distilled* to be a companion book for systems engineers, who need to model a wider spectrum of systems beyond that subset—*software* systems—for which UML was created. With that said, you do not need to know UML as a prerequisite for this book. The structure and content of this book make it a self-sufficient primer for learning SysML.

Structure of the Book

This book contains twelve chapters and two appendixes. Chapter 1, "Overview of Model-Based Systems Engineering," introduces the concept of model-based systems engineering (MBSE) and provides the context and the business case for learning SysML. Chapter 2, "Overview of the Systems Modeling Language," discusses why SysML was created and introduces the nine kinds of SysML diagrams that you can create. Chapter 2 also covers general concepts that apply to all nine kinds of diagrams.

Chapters 3 through 11 zoom in on the details of each of the SysML diagrams, introducing you to the elements and relationships you can display on them. Although there's occasional overlap in the kinds of elements and relationships that can appear on these diagrams, I focus on each diagram one chapter at a time to effectively group related ideas and help you easily locate a particular topic when you need to. Chapters 3–11 are as follows:

- Chapter 3: "Block Definition Diagrams"
- Chapter 4: "Internal Block Diagrams"

- Chapter 5: "Use Case Diagrams"
- Chapter 6: "Activity Diagrams"
- Chapter 7: "Sequence Diagrams"
- Chapter 8: "State Machine Diagrams"
- Chapter 9: "Parametric Diagrams"
- Chapter 10: "Package Diagrams"
- Chapter 11: "Requirements Diagrams"

The last chapter, Chapter 12, "Allocations: Cross-Cutting Relationships," covers the concept of allocations—a relationship that you can use to relate elements across all nine kinds of SysML diagrams.

The sample diagrams in the figures present various aspects of a single system, the DellSat-77 Satellite System—a system I conceived of entirely for the purpose of writing this book (and I hereby certify that I have not disclosed any proprietary information of any aerospace companies). I chose to focus on a satellite system to demonstrate how you can use SysML to model a complex, real-world sociotechnical system—one other than the classic exemplars (ATMs and cruise control systems) that seem to be prevalent in modeling workshops. And I chose to use a single system as a running example threaded through all chapters to show you how the nine kinds of SysML diagrams present complementary and consistent views of an underlying system model.

The SysML model of the DellSat-77 Satellite System is available for download from my website, www.lennydelligatti.com, on the "Articles and Publications" page. I have made the data files available both in XMI format and in the native formats of various modeling tools. This resource enables self-learners as well as instructors and their students to get hands-on with the system model that appears throughout this book in the modeling tool of their choice.

Appendix A, "SysML Notation Desk Reference," is a concise summary of the graphical notations presented in this book, along with references to the sections where they are discussed in detail. Appendix B, "Changes between SysML Versions," covers the kinds of elements that are introduced in SysML v1.3, the latest version of SysML at the time of this writing.

SysML v1.2 is the version of SysML that is currently assessed on the OCSMP certification exams. The biggest differences between SysML v1.2 and v1.3 are in *ports*—a kind of element that can appear on block definition diagrams (BDDs) and internal block diagrams (IBDs). I cover

BDDs in Chapter 3 and IBDs in Chapter 4. In these chapters, I focus on the SysML v1.2 definition of ports for three reasons:

- They are the predominant kinds of ports in system models on modeling projects that started before the release of SysML v1.3—and many of those projects are still active.
- Some modeling tools lag behind the changes in SysML and have not yet implemented the SysML v1.3 definition of ports.
- The OCSMP certification exams have not been revised since the release of SysML v1.3 and still cover the SysML v1.2 definition of ports.

Never fear, though; I give the SysML v1.3 definition of ports full coverage in Appendix B. If your modeling team is about to create a new system model, I recommend using the new kinds of ports instead of the old ones (assuming your SysML modeling tool supports them).

The order of the chapters is loosely based on the typical frequency of use of the diagrams. It does not reflect the relative value of each kind. It can't. Value is a subjective thing. Your team will determine that based on the modeling method you adopt and the deliverables you produce for your customer.

The order of the chapters also does not reflect—and should not suggest—any particular modeling method. Simply put, this is not a methodology book; rather, it's a language book. In Chapter 1, "Overview of Model-Based Systems Engineering," I discuss the distinction between modeling methods and modeling languages. I list a few well-known modeling methods and point to references that discuss them comprehensively.

My goal in this book is to present you with concise, targeted coverage of the most common and most useful features of SysML—features that are useful no matter which modeling method your team adopts. A key point is that SysML is only a language; it's method independent. I designed *SysML Distilled* to be method independent as well. I want you to come away knowing that SysML is a value-added medium for communication no matter which processes, procedures, or tools your team adopts to do your work and meet your stakeholders' needs.

I hope you find this book a valuable companion in your study of SysML. It's a rich, expressive language—one with enough breadth and depth to let you visualize and communicate all aspects of a system's design. There's a lot to know, but you don't need to know all of it to create system models that communicate clearly and effectively. Dive in and get what you need. You'll discover how quickly you can put that knowledge to work and deliver value to your customer.

—Lenny Delligatti
Houston, Texas
October 2013

Acknowledgments

Many talented and dedicated people deserve credit for producing this book. I would like to begin with a special thanks to Jim Thompson—my friend, colleague, spiritual adviser, and partner in weekly sushi catharsis. He spent many months reviewing chapters as I wrote the original manuscript, and he provided valuable and insightful feedback. This book benefited greatly from his keen technical mind and his excellent communication skills.

A special thanks also to Chris Guzikowski at Addison-Wesley. He shepherded this project from its inception and flattened the steep learning curve for this new author. I thank him particularly for the sage advice that got me to the finish line: "Just keep choppin' away at it."

Chris Zahn, development editor at Addison-Wesley, and Betsy Hardinger, copy editor extraordinaire, provided the essential support I needed to hammer this book into its present form. They taught me the art of turning good ideas into good writing and a well-crafted manuscript. The quality of this book is far greater because of their contributions.

Elizabeth Ryan, project editor at Addison-Wesley, coordinated the work of the production team, who created the layout for the book and brought the various pieces together in preparation for printing. They made a complex process look easy and created a polished final product. I'm grateful to them for their hard work.

I'd like to extend my deep appreciation to the exceptional team of engineers and systems modelers who served as technical reviewers for this book: Celso Gonzalez, Robert Cloutier, Susanne Sherba, John Pantone, Michael Engle, and Michael Chonoles. Their expertise and insight enabled me to turn a very rough draft into a significantly more focused final product—one that better serves the systems engineering community. My thanks to all of them.

I would also like to thank the following individuals who graciously agreed to review the revised manuscript and provide endorsements on short notice: Jeff Estefan, Susanne Sherba, Lonnie VanZandt,

Bran Selic, J. D. Baker, Tim Weilkiens, Tom Fargnoli, Robert Cloutier, Matthew Hause, Russell Peak, Doug Tolbert, Celso Gonzalez, and Bobbin Teegarden. These extraordinary people have made significant contributions to the systems modeling community as modeling language developers, modeling certification developers, systems architects, teachers, and expert practitioners in the field. I am honored and humbled to receive their endorsements.

A special thanks to Rick Steiner for writing a foreword for this book. Rick is one of the original creators of SysML, and he continues to serve on the SysML Revision Task Force (RTF)—the team that evolves and improves the SysML specification over time in response to feedback from the systems modeling community. Our profession is richer because of his experience and contributions. My thanks to him for all he's done.

A special thanks also to Richard Soley for his foreword. Richard has led the Object Management Group (OMG) since its inception in 1989. The mark that the OMG has left in the engineering world cannot be overstated. Richard, the OMG staff, and the pantheon of expert engineers who have served as volunteer members of OMG working groups have—without hyperbole—transformed the way we do engineering. The creation of the model-based engineering paradigm and its infusion into the work we do in our profession have enriched those of us who love and practice engineering as well as the customers we serve. My thanks to Richard for his vision and leadership as our community continues to navigate this sea change.

Many brilliant, experienced engineers have contributed to the development of SysML for more than a decade. Any attempt on my part to give each of them due credit individually would fail. I have to resign myself to thanking them collectively for their efforts. It's because of their hard work that we have this rich medium called SysML to communicate our system design ideas to one another. I've strived to make this book a worthy representative of that product.

I am most thankful to my wife, Natalie, and my children, Noelle and Aidan. For two years, they were extraordinarily patient and understanding when I spent long hours on the computer in the evenings and on weekends. They give my life purpose, and I'm deeply grateful to them for their love and support.

About the Author

Lenny Delligatti received his B.S. in electrical and computer engineering from Carnegie Mellon University and his M.S. in computer science systems engineering from the University of Denver. He holds the *OMG Certified Systems Modeling Professional (OCSMP) Model Builder: Advanced* certification, the highest level of certification in SysML and model-based systems engineering (MBSE) methodology. Additionally, he holds the *OMG Certified UML Professional (OCUP): Advanced* certification, the highest level of certification in UML.

Lenny is a senior systems engineer with Lockheed Martin, creating SysML models and serving as the MBSE lead for NASA's Mission Control Center: 21st Century (MCC-21) project at Johnson Space Center. He previously served as an embedded software engineer on NASA's Aircraft Simulation Program (ASP), building VxWorks kernels and developing flight software for NASA's fleet of Gulfstream II in-flight space shuttle simulators. He also served as a software engineer on the Nomad project at Carnegie Mellon University's Field Robotics Center, designing and developing the Sensor Manager subsystem for the Nomad Autonomous Rover.

Lenny is a member of the Object Management Group (OMG) SysML Revision Task Force (RTF) and the OCUP2 Certification Development Team. He also serves as the Education and Outreach Director for the Texas Gulf Coast Chapter (TGCC) of the International Council on Systems Engineering (INCOSE), supporting the professional development of the Houston-area systems engineering community.

In addition to his engineering experience, Lenny served as a Surface Warfare Officer in the U.S. Navy, completing a deployment in support of Operation: Enduring Freedom and two tours of duty in Sasebo, Japan, and Norfolk, Virginia. Following his Navy service, he received formal training in pedagogy at Old Dominion University and earned a license to teach mathematics in the state of Virginia. He served as a mathematics teacher and department head in the Fairfax County public school system before transitioning back into engineering upon his move to Houston, Texas.

Lenny is passionate about engineering and helping engineers develop more effective ways to do engineering. He has created and delivered hundreds of hours of classroom instruction to hundreds of systems and software engineers on the topics of UML, SysML, and MBSE, enabling many to earn OMG certifications and lead MBSE efforts on their projects. He has delivered SysML and MBSE presentations at INCOSE meetings and at American Institute of Aeronautics and Astronautics (AIAA) Technical Symposia at Johnson Space Center.

Chapter 1

Overview of Model-Based Systems Engineering

MBSE is the formalized application of modeling to support system requirements, design, analysis, verification, and validation activities beginning in the conceptual design phase and continuing throughout development and later life cycle phases.

—INCOSE, *Systems Engineering Vision 2020*

You're reading this book to learn the Systems Modeling Language (SysML)—either to create SysML models on your systems engineering team, to earn the OMG (Object Management Group) Certified Systems Modeling Professional (OCSMP) certification, or both. SysML, however, is only one facet of a larger subject: model-based systems engineering (MBSE). MBSE is a practice; it's something you *do*. And SysML is a graphical modeling language that enables you to practice MBSE. The practice of MBSE provides the context—and the business case—for learning SysML.

I begin this chapter by answering the basic question, What is MBSE? I then discuss the three pillars of MBSE—the three enablers for the practice of model-based systems engineering. I end by making you aware of a common myth that has arisen about MBSE so that you can better manage your customers' expectations while you deliver the return on investment that MBSE promises.

1.1 What Is MBSE?

The best way to understand the MBSE approach is to begin by understanding the alternative: what modeling practitioners call the **document-based approach** to engineering—and what nonpractitioners call "the way we've always done things." Whether they apply the document-based approach or MBSE, systems engineers perform the same life cycle activities described in the *INCOSE Systems Engineering Handbook* (INCOSE stands for International Council on Systems Engineering). The key difference between the two approaches, however, is the nature of the primary artifacts produced in those life cycle activities.

With the document-based approach, systems engineers manually generate some subset of the following artifacts: concept of operations (ConOps) documents, requirements specifications, requirement traceability and verification matrices (RTVMs), interface definition documents (IDDs), N^2 charts (also known as N-squared charts—matrices of structural interfaces), architecture description documents (ADDs), system design specifications, test case specifications, and specialty engineering analyses (e.g., analyses of reliability, availability, schedulability, throughput, and response time). Document-based systems engineers produce these artifacts in the form of a disjoint set of text documents, spreadsheets, diagrams, and presentations (and configuration-manage them in a disjoint set of repositories).

The problem is this: The document-based approach to systems engineering is expensive. More precisely, it's more expensive than it needs to be; you incur a significant percentage of total life cycle cost maintaining that disjoint set of artifacts. And if you don't pay that cost, the artifacts become inconsistent and obsolete.

Consider the following day-in-the-life scenario. A system architect makes a fourth-iteration design decision to refactor a single block in the system hierarchy into two blocks to achieve a better separation of concerns. She decides to rename the original block to better convey its new, narrower focus. To implement this change completely and consistently, she needs to locate every text document, table, matrix, diagram, and presentation that contains the block, open each one sequentially from the various file servers, intranet websites, and configuration management (CM) repositories where it resides, and then manually type the same change into all of those artifacts.

This approach is time consuming and error prone. For one thing, the architect may mistype the new block name. More important, she needs to know ahead of time which artifacts will be impacted, perhaps a great many. She will likely miss a handful of them, and they will become inconsistent with the rest of the set. That creates problems for the development teams who rely on them as inputs for their stage of the life cycle. It's also a problem for the project manager, who must account for the schedule slippage and increased life cycle cost to fix any defects that propagate down the line.

This scenario is commonplace in organizations that practice the traditional document-based approach to systems engineering. Inconsistency is the problem. And MBSE—when practiced correctly—is the solution.

With the MBSE approach, systems engineers perform the same life cycle activities and produce the same set of deliverables. But the deliverables are not the immediate outputs of the life cycle activities; they are not the primary artifacts. With the MBSE approach, the primary artifact of those activities is an integrated, coherent, and consistent **system model,** created by using a dedicated systems modeling tool. All other artifacts are secondary—automatically generated from the system model using that same modeling tool.

The system model serves as a central repository for design decisions; each design decision is captured as a **model element** (or a relationship between elements) in a single place within the system model. With the MBSE approach, all diagrams and autogenerated text artifacts are merely *views* of the underlying system model; they are not the model itself. And that distinction is the root of the return on investment (ROI) that MBSE offers over the traditional approach.

Let's return to our day-in-the-life scenario, this time with the MBSE approach. The system architect decides to rename the original (refactored) block to better convey its new, narrower focus. To implement this change completely and consistently, she locates that one block within the system model hierarchy (often by using a keyword search in the modeling tool) and types the new name for that block one time. That's it.

The modeling tool automatically (and instantly) propagates the change to all diagrams where that block appears, no matter how large the set may be. The diagrams, after all, are merely views of the underlying model. If that model changes, the diagrams change. The modeling tool also inserts the change into all the autogenerated text artifacts the next time the architect exports them from the model. There is no opportunity for inconsistency to occur between the various views of the model.

MBSE promises increased quality and affordability for one simple reason: The cheapest defect to fix is the one you prevented. And at the heart of this approach is this new kind of engineering artifact called the system model.

The approach that I've described is actually a hybrid form that bridges the gap between MBSE and the document-based approach. Model-based engineering organizations are constrained to adopt this hybrid approach when their customer requires text artifacts as deliverables for review and approval. Organizations that do not have this constraint, however, can practice MBSE in its pure form.

Organizations that attain the highest MBSE maturity level forgo the creation of text artifacts entirely. The system model itself is the artifact that gets reviewed and approved. It's the artifact that gets handed off, refined, and evolved as it goes from one stage of design to the next.

Organizations that produce software systems can even use a (sufficiently robust) modeling tool to transform a system model into a software model and, ultimately, into production-quality source code. This level of MBSE maturity blurs the line between design and development, enabling rapid prototyping and system simulation. At all times, however, the model remains the primary artifact that gets modified when customer requirements change and new design decisions are made. All other artifacts, including source code, are autogenerated byproducts of the model, continuously consistent with the model and with each other.

This is MBSE.

1.2 The Three Pillars of MBSE

How do you do MBSE? What do you need to know?

In short, you need to know three things: a modeling language, a modeling method, and a modeling tool. I refer to these as the three pillars of MBSE. As a member of a design team that's creating an integrated system model, you will use a dedicated modeling tool to perform a set of design tasks prescribed by a modeling method to add elements (and relationships between elements) to an integrated system model that is expressed in a standard modeling language.

Knowledge of a modeling language alone enables you to sketch system design ideas on paper or a whiteboard to quickly and effectively communicate with other team members. Learning a modeling language is the first skill you should acquire, and it's the focus of this book. However, the practice of MBSE requires you to possess all three skills to gain the ROI that this approach offers.

Let's examine each pillar in more detail.

1.2.1 Modeling Languages

When you create a model, you are speaking a language. It's not the natural language you learned as a child at home and in school. It's not the natural language I'm using to communicate with you right now. Rather, it's a **modeling language:** a semiformal language that defines the kinds of elements you're allowed to put into your model, the allowable relationships between them, and—in the case of a graphical modeling language—the set of notations you can use to display the elements and relationships on diagrams.

MBSE practitioners commonly use the **Systems Modeling Language (SysML)** to construct models of a system's structure, behavior, requirements, and constraints. SysML is the focus of this book, but it's not the only modeling language. Engineers and analysts in other design domains (e.g., systems-of-systems, software, hardware, performance, business processes) have other modeling languages available that are more appropriate for the types of systems they design. Like SysML, some of those languages are graphical modeling languages (e.g., UML, UPDM, BPMN, MARTE, SoaML, IDEFx); others are text modeling languages (e.g., Verilog, Modelica).

The key idea here is that each modeling language is a standardized medium for communication; the rules defined in a given language give the model's elements and relationships unambiguous meaning. The capability to construct and read well-formed models is at the heart of the MBSE approach.

1.2.2 Modeling Methods

Learning a modeling language is only the first step on the MBSE path. A modeling language defines a **grammar:** a set of rules that determines whether a given model is well formed or ill formed. Those rules do not dictate how and when to use the language to create a model; they stop short of dictating any particular modeling method.

In contrast, a **modeling method** is something like a road map; it's a documented set of design tasks that a modeling team performs to create a system model. More precisely, it's a documented set of design tasks that ensures that everyone on the team is building the system model consistently and working toward a common end point. Without such guidance, there will be wide variance in the breadth, depth, and fidelity that each member of the team builds into the system model.

Like all projects, an MBSE project requires a plan. And every plan begins with a purpose. Your team will begin by answering the following questions: Why are you modeling? More precisely, what are the expected results of the modeling effort? Are you creating a model only to serve as the central record of authority for all design decisions? Do you need to autogenerate text artifacts from the model for review and approval? Will you use the model to manage requirements traceability and perform downstream impact analysis? Will you use the model to perform trade studies of alternative configurations? Will the system model be integrated with dedicated equation-solving tools and simulation tools to execute the model directly? Will the model itself be an input for the work of downstream design and development teams, such as software, hardware, reliability/availability/performance analysis? Will the model contain the integration and acceptance test cases that will verify system assembly after development? The answers to these questions determine the **purpose** of your team's modeling effort.

Once your team has defined that purpose, you can then answer a new set of questions. How much of the external environment of your proposed system needs to be modeled? Which parts of your system need to be modeled? Which behaviors need to be modeled? How deeply do you need to decompose the internal structures and behaviors? Which details need to be in the model, and which details can be omitted (and left to the discretion of the development teams at implementation time)? The answers to these questions determine the **scope** of the system model your team needs to build.

The definition of scope sets the goalpost that your team is working toward; it enables your team to determine when the model is complete. To be clear, your team will evolve the model over time as requirements change and new design decisions are made. "Complete" in this context means that the model satisfies the purpose you outlined in the project plan.

The scope of the model also determines the modeling method that your team will follow. Several modeling methods are documented in the literature. Your team can adopt one of those existing methods and tailor it to meet your needs and objectives. Or you can create a custom modeling method if none of the existing ones is a good fit. That discussion, however, is beyond the scope of this book.

The focus here is to help you become proficient in SysML, a modeling language, and not to teach you any particular modeling method. SysML is method independent; you can use SysML to create a system model no matter which modeling method you decide is the best fit for your needs. However, I use a little real estate here to list some well-known modeling methods (and some references that provide in-depth coverage of them) to help you on your journey:

- **Method:** INCOSE Object-Oriented Systems Engineering Method (OOSEM)

  Reference: Friedenthal, Sanford, et al., *A Practical Guide to SysML, Second Edition: The Systems Modeling Language* (Boston: MK/OMG Press, 2011)
- **Method:** Weilkiens System Modeling (SYSMOD) method

  Reference: Weilkiens, Tim, *Systems Engineering with SysML/UML: Modeling, Analysis, Design* (Boston: MK/OMG Press, 2008)
- **Method:** IBM Telelogic Harmony-SE

  Reference: Hoffmann, Hans-Peter, "Harmony-SE/SysML Deskbook: Model-Based Systems Engineering with Rhapsody," Rev. 1.51, Telelogic/I-Logix white paper (Telelogic AB, May 2006)

These modeling methods broadly span many stages of the systems engineering life cycle. Not every step prescribed by these methods will apply to your project. Any modeling method you adopt needs to be tailored to meet your project's specific needs. These methods are good starting points.

1.2.3 Modeling Tools

Developing proficiency with a modeling tool is the third pillar of MBSE. **Modeling tools** are a special class of tools that are designed and implemented to comply with the rules of one or more modeling languages, enabling you to construct well-formed models in those languages.

Modeling tools are distinct from diagramming tools such as Visio, Schematic, SmartDraw, ProcessOn, and others. With a **diagramming tool,** you create diagrams—shapes on a page. There is no model underlying those diagrams that ensures automated consistency between them. In contrast, with a modeling tool, you create a model—a set of elements and relationships between elements, and optionally a set of diagrams that serve as views of the underlying model.

When you modify an element on a diagram within a modeling tool, you're actually modifying the element itself in the underlying model. The modeling tool then instantaneously updates all the other diagrams that display that same element. This is a powerful capability—and one that is offered only by this distinct class of tools.

Note that a modeling language specification, such as SysML, is vendor neutral. A particular modeling tool is one vendor's implementation of that language specification. Several commercial tool vendors and nonprofit consortiums have created modeling tools for the various modeling languages. These tools vary in cost, capability, and compliance with the modeling language specifications. Selecting the best tool—based on your project's specific needs and cost constraints—should be part of the MBSE adoption process in your organization.

Much like SysML and the other modeling languages, I am vendor neutral. I do not hawk one product over another in this book. However, I list some SysML modeling tools for you to research for the trade study that your organization will inevitably conduct. The following are commercial-grade (euphemism for "not free") modeling tools:

- Agilian (vendor: Visual Paradigm)
- Artisan Studio (vendor: Atego)
- Enterprise Architect (vendor: Sparx Systems)
- Cameo Systems Modeler (vendor: No Magic)
- Rhapsody (vendor: IBM Rational)
- UModel (vendor: Altova)

The following are free modeling tools, offered with an Eclipse Public License (EPL) or General Public License (GPL):

- Modelio (creator: Modeliosoft)
- Papyrus (creator: Atos Origin)

There are many factors to consider when you select a tool. However, I strongly recommend that you select a tool that is XML Metadata

Interchange (XMI) compliant. The XMI standard enables compliant tools to exchange model data. This capability will ensure that your team avoids the vendor lock-in trap when your needs (and cost constraints) change in the future.

1.3 The Myth of MBSE

The myth of MBSE arises among **stakeholders**—external customers and internal downstream design and development teams—who know *about* MBSE but don't practice it themselves. These are the stakeholders who expect deliverables from you. They understand—conceptually, at least—that you will autogenerate those deliverables from the system model.

The myth they harbor deep inside is that MBSE is an Easy Button: You push it, and good things pop out. To put this more concretely, they believe, incorrectly, that MBSE makes every engineering task easier and reduces cost at every point in the life cycle.

But in truth, MBSE does not (and cannot) eliminate the difficult work of architecting and designing a system well. It does not eliminate the need for engineering rigor during system specification and design—the same rigor that has always been necessary to produce any successful system.

Modeling well is difficult. Designing well is difficult. It's possible to create a bad model. And it's possible to create a good model of a poorly designed system. Creating a good model of a well-designed system to the degree of breadth, depth, and fidelity required to satisfy the model's purpose takes time and hard work and discipline. Simply put, you can't get good things out of the model unless you do the hard work of putting them in there in the first place.

MBSE delivers its ROI when change happens—when new design decisions are made and stakeholders' needs evolve throughout the system life cycle. And change, of course, inevitably happens. Until that time, manage your stakeholders' expectations and dispel the myth of MBSE when it presents itself.

Summary

MBSE is an approach to performing systems engineering that promises to deliver a greater return on investment than the traditional document-based approach. The practice of MBSE rests on three pillars: a modeling language, a modeling method, and a modeling tool. The chapters that follow help you put that first pillar into place by teaching you SysML, the graphical modeling language that has become the de facto standard among MBSE practitioners.

Chapter 2

Overview of the Systems Modeling Language

SysML is a broad and richly expressive graphical modeling language, enabling you to visualize and communicate the essential aspects of a system's design: structure, behavior, requirements, and parametrics (mathematical models). SysML can serve as the first of the three pillars of MBSE, as discussed in Chapter 1, "Overview of Model-Based Systems Engineering."

This chapter provides a high-level view of SysML: the purpose of SysML as a whole, the purpose of each of the nine kinds of SysML diagrams, and overarching concepts that apply to all of them. This discussion provides important context for the in-depth coverage of each kind of diagram in the chapters that follow.

2.1 What SysML Is—and Isn't

SysML is one of several graphical modeling languages. The key word is *language*. SysML is a language—a medium for communicating ideas from one person to another. It has a grammar and a vocabulary just like any of the natural languages we speak (e.g., Hindi, Japanese, English). SysML is the language "spoken" by MBSE practitioners when they create system models to visualize and communicate ideas about their systems' designs to other stakeholders.

I put "spoken" in quotes because SysML is a **graphical language**. Its vocabulary consists of graphical notations that have specific meanings. For example, an arrow with a dashed line has a different meaning from an arrow with a solid line (more on these details in the chapters that follow). The key point is the purpose of SysML: visualization and communication of a system's design among stakeholders.

The grammar and notations of SysML are defined in a standards specification that is owned and published by the Object Management Group, Inc. (OMG). The OMG is a consortium of hundreds of computer industry companies, government agencies, and academic institutions that collaborate to develop a set of enterprise integration standards and promote business technology. You can find out more about the OMG on its website: www.omg.org. You will find links there also to the SysML specification document—the primary source of information about the rules of SysML.

The SysML specification attempts to define the grammar of this modeling language in a way that is precise and unambiguous, and it largely succeeds. But the text in the specification can sometimes be difficult to parse and process. The target audiences for this primary source of SysML information are the vendors of modeling tools and designers of modeling languages (and also modeling geeks who write books on the subject). In short, the SysML specification is not meant for beginners.

If that's not daunting enough, it's only half the story. As I discuss in the next section, SysML is not an independent, stand-alone language. Rather, it's a **profile**—an **extension**—of a subset of the Unified Modeling Language (UML). Therefore, if you wanted the complete definition of the grammar and vocabulary of SysML, you would need to refer also to parts of the UML specification document (also available on the OMG website).

If you're new to SysML, I recommend that you flip to Appendix A, "SysML Notation Desk Reference," to get a good first look at its various graphical notations. These notations are what actually appear on the SysML diagrams you create. These notations form the vocabulary of the systems modeling language.

So that's what SysML is: It's a modeling language. It's equally important to understand what SysML is not: It's not a modeling method. To be clear, the formal SysML specification document defines only the language itself (its grammar and vocabulary); it does not prescribe any particular modeling method.

Note

Please refer to Chapter 1 for a discussion of the distinction between modeling languages and modeling methods.

The SysML specification does not tell you, for example, at which point in the life cycle you should create, say, a use case diagram. It does not dictate that you must use an activity diagram to elaborate a use case. It does not demand that you create a set of internal block diagrams (IBDs), each focusing on a particular aspect of the system architecture. All these are methodological decisions. They fall outside the scope of the SysML specification. It is up to you and your project team to adopt and tailor a modeling method that will enable you to achieve your project's unique objectives.

2.2 Yes, SysML Is Based on UML—but You Can Start with SysML

As mentioned earlier, SysML is not an independent language; it's a profile, or extension, of UML created specifically for the systems engineering domain. UML was designed to be a standard modeling language for the software engineering domain. Systems engineers saw value in the practice of using a standard modeling language to construct models of systems, but they didn't feel that UML sufficiently captured all the concepts that are meaningful in systems engineering.

For example, UML models can contain *DataType* elements. Software engineers can use a data type (e.g., *Integer*) in a UML model to specify the type of an attribute within a class, the type of an object that can flow through an activity, and the type of a parameter within an operation. Systems engineers, however, care about other types of things that can flow—matter and energy—and not just data. The concept of *DataType* simply wasn't sufficient. Therefore, SysML introduces a new kind of model element called *ValueType*, which extends the concept of *DataType* to provide a more neutral term for the broader set of types in the systems engineering domain.

Because SysML is an extension of UML, some of the rules of SysML are actually defined in the UML specification document. This means that the SysML specification document is not a self-sufficient definition of the language. For example, if you wanted to know all the rules about using a value type in a system model, it wouldn't be sufficient to read the *ValueType* entry in the SysML specification; you would have to read the *DataType* entry in the UML specification, too. That's simply the nature of a profile as a language extension mechanism.

So do you need to go buy a UML book to learn how to create a system model?

No, you don't. This book is a sufficient primer for learning SysML so that you can get started modeling as quickly as possible. When I discuss value types in Section 3.9, "Value Types," I tell you everything you need to know about them to use them correctly and effectively in a system model. It will be transparent to you whether a specific detail comes from the UML definition of the base element, *DataType*, or from the SysML extension, *ValueType*.

2.3 SysML Diagram Overview

There are nine kinds of SysML diagrams:

- Block definition diagram (BDD)
- Internal block diagram (IBD)
- Use case diagram
- Activity diagram
- Sequence diagram
- State machine diagram
- Parametric diagram
- Package diagram
- Requirements diagram

Figure 2.1, which appears in the SysML specification v1.2, provides a good overview of the categories and relationships between the SysML diagrams. But this figure is meaningful only if you know what the lines

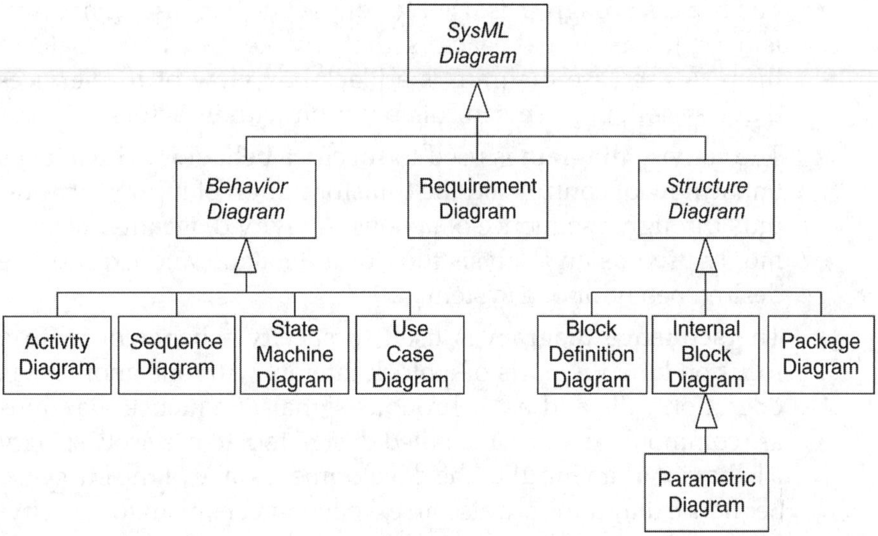

Figure 2.1 *SysML diagram taxonomy*

with the hollow, triangular arrowheads mean. They're called *generalizations*. You read them as "is a type of" in the direction of the arrowhead. (I discuss generalizations in detail in Section 3.6, "Generalizations.")

With this in mind, Figure 2.1 conveys quite a bit of information. Activity diagrams, sequence diagrams, state machine diagrams, and use case diagrams are types of **behavior** diagrams. Block definition diagrams, internal block diagrams, and package diagrams are types of **structure** diagrams. Parametric diagrams are a type of internal block diagram; therefore, a parametric diagram is transitively a type of structure diagram. Finally, **requirements** diagrams are in a category by themselves—but still a useful addition to this family of SysML diagrams.

Here's a brief summary of the purpose of each kind of diagram.

- The **block definition diagram** (BDD) is used to display elements such as blocks and value types (elements that define the types of things that can exist in an operational system) and the relationships between those elements. Common uses for a BDD include displaying system hierarchy trees and classification trees.
- The **internal block diagram** (IBD) is used to specify the internal structure of a single block. More precisely, an IBD shows the connections between the internal parts of a block and the interfaces between them.

- The **use case diagram** is used to convey the use cases that a system performs and the actors that invoke and participate in them. A use case diagram is a black-box view of the services that a system performs in collaboration with its actors.
- The **activity diagram** is used to specify a behavior, with a focus on the flow of control and the transformation of inputs into outputs through a sequence of actions. Activity diagrams are commonly used as an analysis tool to understand and express the desired behavior of a system.
- The **sequence diagram** is used to specify a behavior, with a focus on how the parts of a block interact with one another via operation calls and asynchronous signals. Sequence diagrams are commonly used as a detailed design tool to precisely specify a behavior as an input to the development stage of the life cycle. Sequence diagrams are also an excellent mechanism for specifying test cases.
- The **state machine diagram** is used to specify a behavior, with a focus on the set of states of a block and the possible transitions between those states in response to event occurrences. A state machine diagram, like a sequence diagram, is a precise specification of a block's behavior that can serve as an input to the development stage of the life cycle.
- The **parametric diagram** is used to express how one or more constraints—specifically, equations and inequalities—are bound to the properties of a system. Parametric diagrams support engineering analyses, including performance, reliability, availability, power, mass, and cost. Parametric diagrams can also be used to support trade studies of candidate physical architectures.
- The **package diagram** is used to display the way a model is organized in the form of a package containment hierarchy. A package diagram may also show the model elements that packages contain and the dependencies between packages and their contained model elements.
- The **requirements diagram** is used to display text-based requirements, the relationships between requirements (containment, derive requirement, and copy), and the relationships between requirements and the other model elements that satisfy, verify, and refine them.

2.4 General Diagram Concepts

You should be aware of a few overarching concepts about SysML diagrams before you delve into the details of the specific types. A sample SysML diagram is shown in Figure 2.2.

Each diagram has a frame, a contents area (colloquially called the "canvas"), and a header. The diagram **frame** is the outer rectangle. The **contents area** is the region inside the frame where model elements and relationships can be displayed. The **header** is in the upper-left corner of the diagram, shown in Figure 2.2 with its lower-right corner cut off.

In SysML (unlike in UML), the frame must be displayed. With that said, some figures in this book show model elements and relationships without an enclosing frame. I do that to hide inconsequential information and focus your attention on particular notations. Officially, though, the frame is mandatory.

One of the most important diagram concepts is the format of the header information. The header commonly contains four pieces of information:

- Diagram kind
- Model element type
- Model element name
- Diagram name

The format of that information is shown in the header in Figure 2.3.

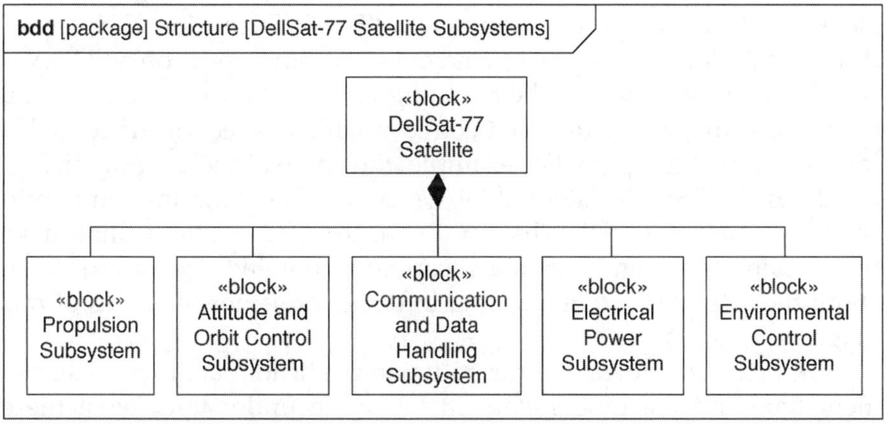

Figure 2.2 *Sample SysML diagram*

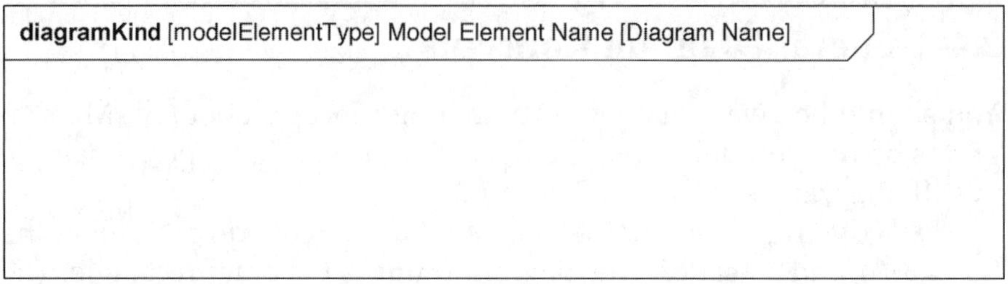

Figure 2.3 *Diagram header format*

I begin with two intuitive pieces of information: diagram kind and diagram name. The **diagram kind** is shown as its SysML-defined abbreviation:

- **bdd** = block definition diagram
- **ibd** = internal block diagram
- **uc** = use case diagram
- **act** = activity diagram
- **sd** = sequence diagram
- **stm** = state machine diagram
- **par** = parametric diagram
- **req** = requirements diagram
- **pkg** = package diagram

Based on this, you can conclude that the diagram in Figure 2.2 is a block definition diagram. (Chapter 3, "Block Definition Diagrams," discusses in detail the kinds of elements that can appear on a BDD.)

The **diagram name** can be anything you want it to be. I advise you to choose a diagram name that conveys which aspect of the model is in focus on that diagram. For example, the name of the diagram in Figure 2.2 is "DellSat-77 Satellite Subsystems." This diagram name indicates that the focus of the diagram is the set of subsystems that make up the satellite system. The model certainly contains other information about the satellite system, but that information is not the focus of this diagram.

The next two pieces of information in the header are the model element type and the model element name. To understand what these refer to, you first need to know another essential concept about SysML diagrams: Each diagram you create represents an element that you've defined somewhere in your system model. More precisely, the diagram frame represents an element in the model. And *that* model element is the element whose type and name appear in the diagram header.

In Figure 2.2, the model element type is "package," and the model element name is "Structure." This conveys that the frame of this BDD represents the *Structure* package that exists somewhere in the system model hierarchy.

Requiring each diagram to represent a model element may seem like a strict and unnecessary constraint, but in the words of Frederick Brooks, Jr., in *The Design of Design* (Boston: Addison-Wesley, 2010), "Constraints are friends" (p. 127). The connection between a diagram and a model element was a deliberate and brilliant decision on the part of the SysML authors. The reason is conveyed by the next key concept of SysML diagrams: The model element represented by the diagram defines the **namespace**—the container element within the model hierarchy—for the other elements shown on the diagram. Simply put, the model element type and model element name shown in the diagram header indicate where the elements on the diagram can be found within the model.

The diagram header in Figure 2.2 tells us that the six blocks shown in the contents area are contained in (i.e., nested under) the *Structure* package in the model hierarchy. This gives us a sense of how elements are partitioned in the model and aids us in navigating it.

The model element may be a structural element (e.g., a package or block), or it may be a behavioral element (e.g., an activity, interaction, or state machine). The type of model element that a diagram can represent depends on the kind of diagram you're creating. The pairings are shown in Table 2.1.

Recall from Chapter 1 the distinction between a system model (as an engineering artifact in its own right) and the set of diagrams you create (which are views of the underlying model). This idea is so important that here I rephrase it more formally and give it a name: the fundamental precept of model-based engineering. Your assignment is to assume full lotus position and repeat the following mantra until you enter a deep meditative state:

> ***A diagram of the model is never the model itself; it is merely one view of the model.***

This idea is a paradigm shift for engineers who have only ever sketched designs using paper, whiteboards, or diagramming tools. And it's an idea best explained via metaphor: The model is a mountain,

Table 2.1 *Allowable Model Element Types for Each Diagram Kind*

Diagram Kind	Allowable Model Element Types
Block definition diagram	package, model, modelLibrary, view, block, constraintBlock
Internal block diagram	block
Use case diagram	package, model, modelLibrary, view
Activity diagram	activity
Sequence diagram	interaction
State machine diagram	stateMachine
Parametric diagram	block, constraintBlock
Requirement diagram	package, model, modelLibrary, view, requirement
Package diagram	package, model, modelLibrary, view, profile

and a diagram is a picture of the mountain. The mountain exists whether or not anyone takes a picture of it. If a man comes along and takes a picture from the north side, he creates one view of the mountain that shows some of its features but not others. If a woman takes a picture from the west side, she creates a second view of the mountain that shows a different—possibly overlapping—set of features.

Each of the two views focuses on a different aspect of the whole. But at all times the pictures are only views of the mountain and not the mountain itself. The mountain and its features will continue to exist even if the photographers later crop out certain details from the final pictures . . . and even if they destroy the pictures.

This metaphor fails in one respect: You can add new features and modify or delete existing features from the model at any time. When you modify an existing feature, the changes are instantly reflected on all diagrams that show the feature. When you delete a feature from the model, it instantly vanishes from all diagrams that showed that feature. Clearly the pictures in your photo album don't offer the same capability.

Earlier in this section I mentioned the practice of eliding inconsequential information on a given diagram. No diagram should attempt to convey every detail; the diagram would be unreadable. You should instead decide what you want the focus of a given diagram to be and then elide all model information that is not within that focus. This idea leads to the corollary to the fundamental precept of model-based engineering:

> ***You cannot conclude that a feature doesn't exist from its absence on a diagram; it may be shown on another diagram of the model or on no diagram at all.***

Summary

SysML is a richly expressive graphical modeling language that you can use to visualize the structure, behavior, requirements, and parametrics of a system and communicate that information to others. SysML defines nine kinds of diagrams that you can use to convey all this system design information; each kind serves a specific purpose and conveys specific information about an aspect of a system.

The chapters that follow provide detailed coverage of these diagrams. You will learn the various kinds of SysML model elements—and the relationships among them—that can appear on each kind of diagram. I include discussions of the rules of SysML that you will need to know to build your system model correctly and ensure effective communication with your stakeholders.

Chapter 3

Block Definition Diagrams

The most common kind of SysML diagram is the block definition diagram. You can display various kinds of model elements and relationships on a BDD to express information about a system's structure. You can also adopt design techniques for creating extensible system structures, a practice that reduces the time and cost to change your design as your stakeholders' needs evolve.

3.1 Purpose

The model elements that you display on BDDs—blocks, actors, value types, constraint blocks, flow specifications, and interfaces—serve as types for the other model elements that appear on the other eight kinds of SysML diagrams. We refer to elements that appear on BDDs as **elements of definition**. Elements of definition, in a real sense, form the foundation for everything else in your system model. That's why I'm covering BDDs first.

Elements of definition are important; the structural relationships among them—associations, generalizations, and dependencies—are arguably more important. You display these relationships on BDDs, too. With these relationships, you often create BDDs that convey system decomposition and type classification.

3.2 When Should You Create a BDD?

Often. You should create a BDD often.

That may seem like a glib answer, but it's accurate. BDDs are not tied to any particular stage of the system life cycle or level of design. You and your team will create them (and refer to them) when you perform all the following systems engineering activities: stakeholder needs analysis, requirements definition, architectural design, performance analysis, test case development, and integration. And you often create a BDD in conjunction with other SysML diagrams to provide a complementary view of an aspect of your system of interest.

In short, you should—and will—create BDDs often.

3.3 The BDD Frame

The diagram kind abbreviation for a block definition diagram is *bdd*. The model element type that the diagram frame represents can be any of the following:

- *package*
- *model*
- *modelLibrary*
- *view*
- *block*
- *constraintBlock*

As discussed in Section 2.4, "General Diagram Concepts," the model element that the diagram represents serves as the namespace for the other elements shown on the diagram. A **namespace** is simply a model element that's allowed to contain other model elements; that is, it can have other elements nested under it within the model hierarchy. A namespace, therefore, is a concept that has meaning only within your system model; it has no meaning within an instance of your system.

Many kinds of SysML elements can serve as namespaces. A **package,** however, is the most common kind of namespace for the various elements of definition that appear on BDDs. Therefore, the element that's named in the header of a BDD typically is a package you've created somewhere in the model hierarchy.

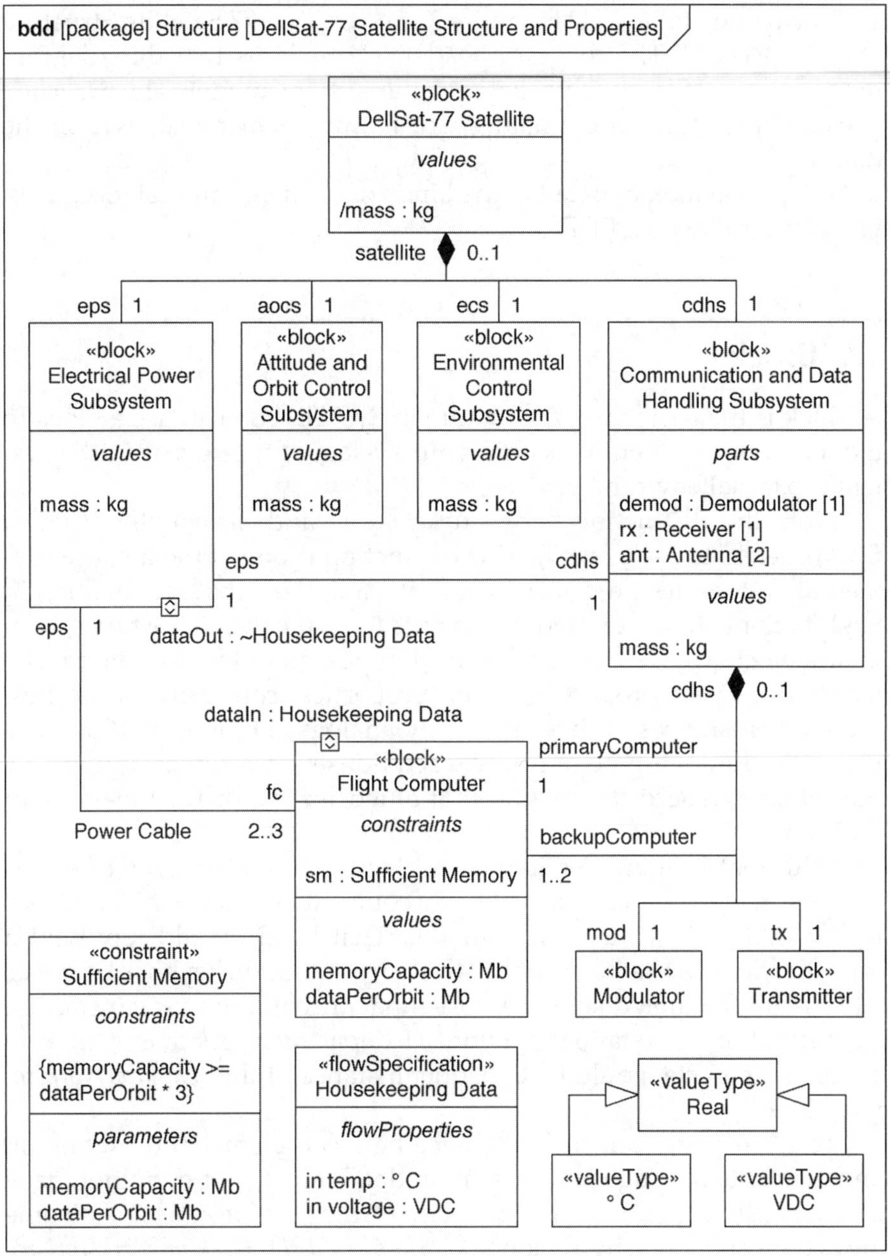

Figure 3.1 *A sample block definition diagram (BDD)*

The name of the BDD in Figure 3.1 is "DellSat-77 Satellite Structure and Properties." The diagram header also tells us that this diagram represents the *Structure* package in the system model. The *Structure* package, therefore, is the namespace for the elements shown on the diagram.

Let's take a look in detail at the kinds of elements and relationships you can display on a BDD.

3.4 Blocks

A **block** is the basic unit of structure in SysML. You can use a block to model any type of entity within your system of interest or in the system's external environment.

Note the distinction between *definition* and *instantiation* (which SysML refers to as "usage"). This distinction is one of the most fundamental system design concepts, and it's a pattern that recurs often in SysML. Some kinds of model elements (e.g., blocks, value types, constraint blocks) represent **definitions** of types; other kinds of model elements (e.g., part properties, value properties, constraint properties) represent **instances** of those types. By analogy, a blueprint of a house is a definition of a type of house; each house a developer builds on a plot of land in accordance with that blueprint is a distinct instance of that type.

With that in mind, I reiterate: A block represents a type of entity, and not an instance. For example, you could create a block named *DesktopWorkstation* in your system model. That block would represent a type that defines a set of properties—such as monitor, keyboard, mouse, CPU, manufacturer, disk space, cost—that are common to all instances. Each desktop workstation that your IT department purchases for each office and cubicle would be a distinct instance of that *DesktopWorkstation* block.

You can easily tell the difference between elements of definition and elements of usage in a system model. Elements of definition have a name only (e.g., *DesktopWorkstation*); elements of usage have a name and a type, separated by a colon (e.g., *SDX1205LJD : DesktopWorkstation*).

The notation for a block is a rectangle with the stereotype «block» preceding the name in the name compartment (as shown in Figure 3.2). You're required to display a block's name compartment. Often you'll display additional optional compartments that convey the features of the block.

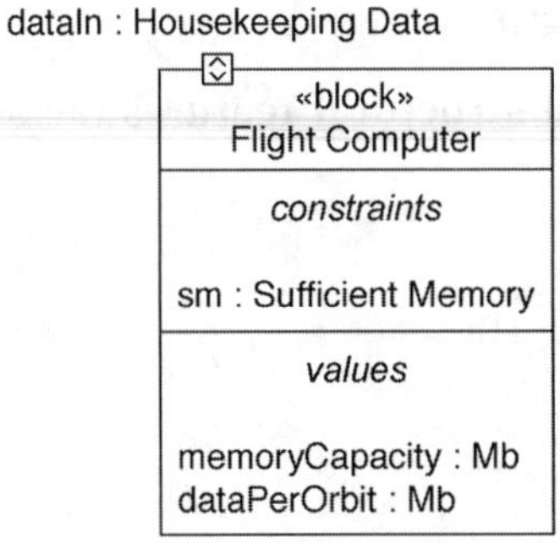

Figure 3.2 *A block*

Features come in two varieties: structural features (also known as properties) and behavioral features. I discuss each category in depth in the next two sections.

Here are the optional compartments that you can display:

- Parts
- References
- Values
- Constraints
- Operations
- Receptions
- Standard ports (in SysML v1.2 and earlier)
- Flow ports (in SysML v1.2 and earlier)
- Full ports (in SysML v1.3)
- Proxy ports (in SysML v1.3)
- Flow properties (in SysML v1.3)
- Structure

The structure compartment is the only compartment that doesn't list features. Rather, it's a graphical compartment that displays a block's internal structure; you can display in that compartment all the same notations you can display on an internal block diagram (IBD). Modelers rarely display this compartment.

Note that even though it's legal to display a block's ports in compartments, it's much more common to display ports as small squares that straddle the border of a block (as shown in Figure 3.2). I discuss ports in detail in Section 3.4.1.5, "Ports."

3.4.1 Structural Features

There are five kinds of **structural features** (also known as **properties**) that a block can own:

- Part properties
- Reference properties
- Value properties
- Constraint properties
- Ports

3.4.1.1 Part Properties

Part properties are listed in the parts compartment of a block (as shown in Figure 3.3). A **part property** represents a structure that's internal to a block. Stated differently, a block is *composed of* its part properties. This relationship conveys ownership.

However, SysML stops short of defining the word *ownership;* this concept has different meanings in different domains. In the hardware domain, ownership typically refers to physical composition. For example, Figure 3.3 conveys that a valid instance of the *Communication and Data Handling Subsystem* block is one that is physically composed of the required parts: flight computers, modulator, demodulator, transmitter, receiver, and antennas. In the software domain, however, ownership

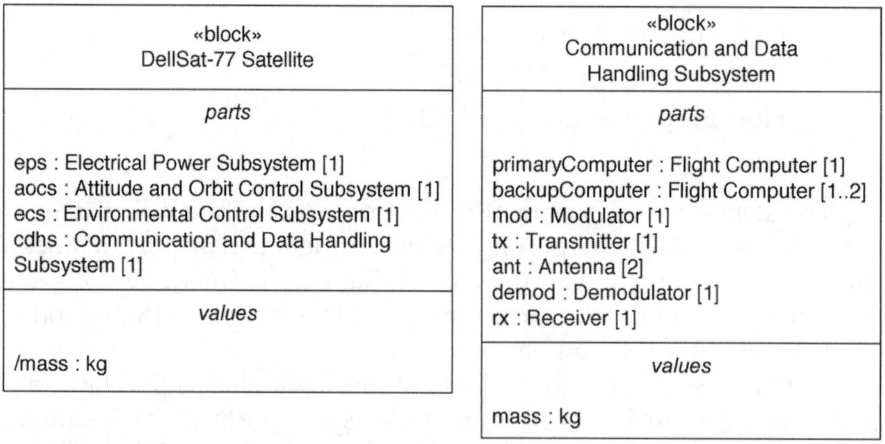

Figure 3.3 *Blocks with part properties*

typically refers to one object's responsibility for the creation and destruction of another object. When memory is allocated for a composite object, memory is allocated for each of its parts, too; similarly, when memory is freed for a composite object, memory is also freed for each of its parts.

But SysML states definitively that *ownership* means that a part property can belong to only one composite structure at a time. However, a part property can be removed from one instance of a composite structure and added to another. For example, I can install a given antenna on only one satellite at a time, and not on two or more simultaneously. But that antenna can be removed from one satellite and reinstalled on another at some point.

When you list a part property in the parts compartment of a block, it appears as a string with the following format:

```
<part name> : <type> [<multiplicity>]
```

The part name is modeler defined. The type generally is the name of a block that you've created somewhere in the system model. The multiplicity is a constraint on the number of instances that the part property can represent within the composite, expressed either as a single integer or as a range of integers.

For example, Figure 3.3 conveys that a valid instance of the *Communication and Data Handling Subsystem* block must be composed of exactly one instance of the *Flight Computer* block—an instance that serves in the role of *primaryComputer.* Additionally, it must be composed of either one or two more instances of *Flight Computer*—instances that serve in the role of *backupComputer.*

If you want a part property to represent an unconstrained number of instances, you can set the multiplicity to 0..*. The asterisk means that there's no upper bound (or more precisely, that you're not specifying an upper bound in the system model). You would read 0..* in English as "zero or more." Alternatively, you can set the multiplicity to *, a shorthand notation for 0..*.

If no multiplicity is shown for a part property, the default is 1 (which is equivalent to 1..1). Note that 1 is almost always the default multiplicity in SysML. There is an important exception, however, which I discuss in Section 3.5.2, "Composite Associations."

When a part property has a multiplicity with an upper bound greater than 1 (e.g., 1..2, 0..10, *), we refer to that part property as a **collection** (of instances). The key idea is that *part property* and *instance* are not synonyms; a single part property may potentially represent multiple instances within a composite if its specified multiplicity allows it.

3.4.1.2 *Reference Properties*

Reference properties are listed in the references compartment of a block (as shown in Figure 3.4). A **reference property** represents a structure that's external to a block.

Unlike a part property, a reference property does not convey ownership. A reference property can roughly be described as a "needs" relationship; a block with a reference property needs that external structure for some purpose, either to provide a service or to exchange matter, energy, or data. And this implies that some type of connection must exist between them.

Note that the presence of a reference property in a block does not by itself convey its purpose. If you need to convey that purpose, you could do so on an internal block diagram (IBD). I discuss this more in Chapter 4, "Internal Block Diagrams."

When you list a reference property in the references compartment of a block, it appears as a string with the following format:

```
<reference name> : <type> [<multiplicity>]
```

The reference name is modeler defined. The type must be the name of a block or actor that you've created somewhere in the system model. The multiplicity is a constraint on the number of instances that the reference property can represent.

For example, Figure 3.4 shows that the *Electrical Power Subsystem* block has a reference property named *cdhs*. This model conveys that an

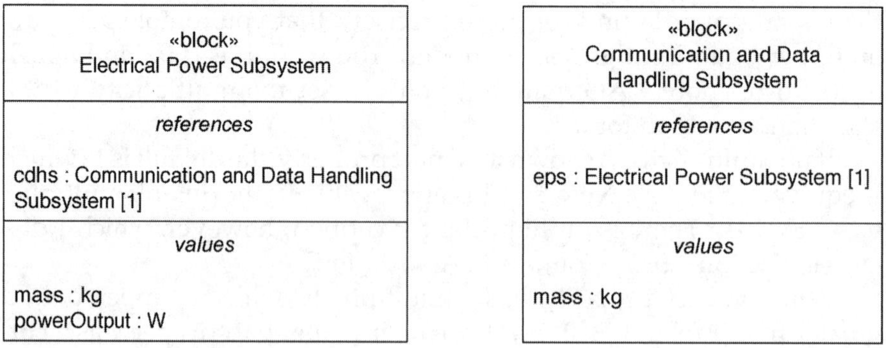

Figure 3.4 *Blocks with reference properties*

instance of *Electrical Power Subsystem* needs exactly one instance of *Communication and Data Handling Subsystem* (to fulfill its design purpose). Again, this view alone doesn't convey what that purpose is; it simply conveys that some type of connection must exist between them.

Like a part property, a reference property's default multiplicity is 1 (if no multiplicity is shown). And like a part property, a reference property is referred to as a collection when its multiplicity has an upper bound greater than 1.

3.4.1.3 Value Properties

Value properties are listed in the values compartment of a block (as shown in Figure 3.5). A **value property** can represent a quantity (of some type), a Boolean, or a string. Most often, though, a value property is something you can assign a number to. Value properties are particularly useful in conjunction with constraint properties to construct a mathematical model of your system (more on this in Chapter 9, "Parametric Diagrams").

When you list a value property in the values compartment of a block, it appears as a string with the following format:

```
<value name> : <type> [<multiplicity>] = <default value>
```

The value name is modeler defined. The type must be the name of a value type that you've created somewhere in the system model. The multiplicity is a constraint on the number of values that the value property can hold. The default value is an optional piece of information; it

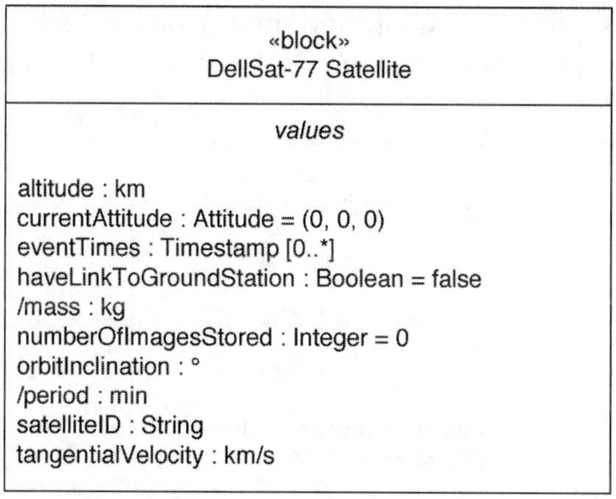

Figure 3.5 *A block with value properties*

represents the value assigned to the value property when an instance of its owning block first gets created.

Figure 3.5 shows that the *DellSat-77 Satellite* block has several value properties. The *eventTimes* value property can hold an unconstrained number of *Timestamp* values (as conveyed by the multiplicity 0..*). *Timestamp* is a value type that exists somewhere in the model hierarchy (more on value types in Section 3.9, "Value Types").

As with a part property and a reference property, a value property's default multiplicity is 1 (if no multiplicity is shown). Similarly, a value property is referred to as a collection when its multiplicity has an upper bound greater than 1.

Some value properties hold values that are assigned, and others hold values that are derived (calculated) from other value properties in the system model. To convey that a value property is derived, you put a forward slash (/) in front of its name. For example, Figure 3.5 shows that the *DellSat-77 Satellite* block owns two derived value properties: *mass* and *period*. This view of the model does not convey the equations used to calculate those derived values, nor does it show which other value properties provide inputs for those equations. You would specify those mathematical relationships using constraint expressions, as discussed in the next section.

3.4.1.4 Constraint Properties

Constraint properties are listed in the constraints compartment of a block (as shown in Figure 3.6). A **constraint property** generally represents a mathematical relationship (an equation or inequality) that is imposed on a set of value properties. This is a higher level of model fidelity than is required on most modeling projects. However, constraint properties are an essential part of constructing mathematical models of

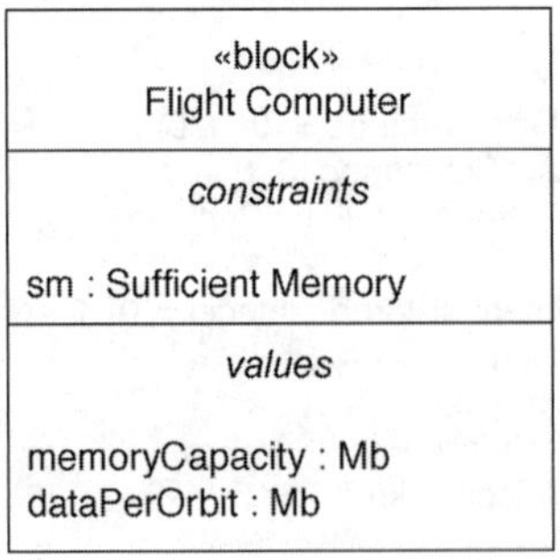

Figure 3.6 *A block with a constraint property*

a system, which you display on parametric diagrams (more on this in Chapter 9).

When you list a constraint property in the constraints compartment of a block, it appears as a string with the following format:

```
<constraint name> : <type>
```

The constraint name is modeler defined. The type must be the name of a constraint block that you've created somewhere in the system model.

A constraint block is simply a special kind of block—one that you create to encapsulate a reusable **constraint expression**. Most often, a constraint expression is an equation or an inequality. For example, Figure 3.7 shows a constraint block named *Sufficient Memory*, which encapsulates the constraint expression

```
memoryCapacity >= dataPerOrbit * 3
```

This constraint block serves as the type for the constraint property *sm* in the *Flight Computer* block (shown in Figure 3.6). This conveys that the values held in the two value properties (*memoryCapacity* and *dataPerOrbit)* must satisfy that mathematical relationship at all times (in a system that's operating nominally).

Note that you're not required to use constraint blocks to impose mathematical relationships on value properties. It's perfectly legal to specify a constraint expression directly in the constraints compartment of a block (as shown in Figure 3.8). You would do this when only one block needs that constraint expression (i.e., when you don't intend to

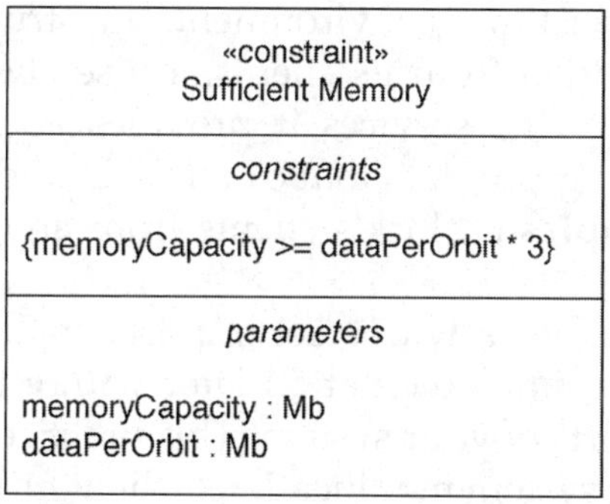

Figure 3.7 *A constraint block*

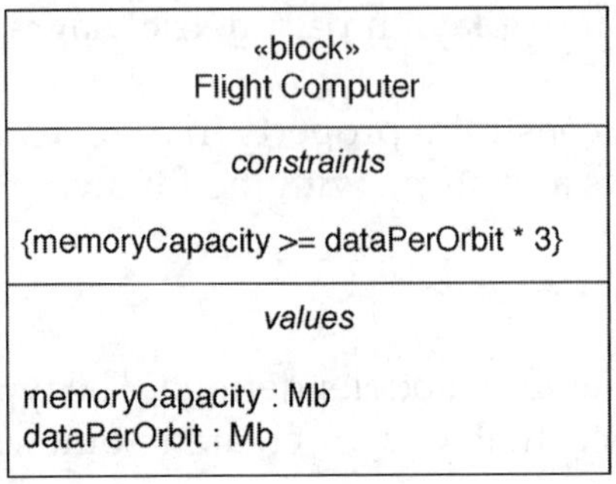

Figure 3.8 *A block with a (non-reusable) constraint*

reuse it in multiple places). As a matter of best practice, though, I recommend that you always encapsulate equations and inequalities in constraint blocks; it enables reuse if the need arises.

I discuss constraint blocks in greater detail in Section 3.10, "Constraint Blocks." Meanwhile, keep in mind these key ideas:

- Blocks can own constraint properties (to constrain value properties).
- Constraint properties are typed by constraint blocks, which generally encapsulate mathematical relationships.

3.4.1.5 Ports

A **port** is a kind of property that represents a distinct interaction point at the boundary of a structure through which external entities can interact with that structure—either to provide or request a service or to exchange matter, energy, or data.

When you add a port to a block, you're modeling a structure as a black box with respect to its environment; the structure's internal implementation is hidden from its clients. Those clients know only the structure's interface (the services it provides and requires, and the types of matter, energy, or data that can flow in and out). Stated differently, a port **decouples** a block's clients from any particular internal implementation.

Encapsulating a block with a set of ports enables you to redesign that block's internal implementation later without impacting the design of the other parts of your system. This practice reduces the time it takes to implement system modifications when the customer's requirements change later in the life cycle, and time saving translates into cost saving.

A port can represent any type of interaction point you need to model. For example, it can represent a physical object on the boundary of a hardware object (e.g., a spigot, an HDMI jack, a fuel nozzle, a gauge). It can represent an interaction point on the boundary of a software object (e.g., a TCP/IP socket, a message queue, a shared memory segment, a graphical user interface, a data file). And it can represent an interaction point between two business organizations (e.g., a purchase order, a courier, a website, a mailbox). SysML imposes no constraints on what a port can represent.

SysML v1.2 (and earlier) defines two kinds of ports—standard ports and flow ports—that you can add to a block to specify different aspects of its interface. A **standard port** lets you specify an interaction point with a focus on the services that a block provides or requires; a **flow port** lets you specify an interaction point with a focus on the types of matter, energy, or data that can flow in and out of a block.

Note

SysML v1.3 no longer supports standard ports and flow ports, instead defining two new kinds of ports: full ports and proxy ports. I discuss these in detail in Appendix B. I focus on standard ports and flow ports in this chapter because they continue to be the predominant kinds of ports in system models at the time of this writing. Additionally, the current versions of the OCSMP certification exams cover the concepts of standard ports and flow ports. Moreover, some modeling tools continue to lag behind the changes in SysML and do not yet support full ports and proxy ports.

Standard Ports A standard port models the services (behaviors) that a block provides or requires at an interaction point on its boundary. Most often, you display a standard port as a small square straddling the border of a block (as shown in Figure 3.9). Note that it's legal to list a standard port as a string in the standard ports compartment, but this is an uncommon notation.

A standard port can have a modeler-defined name (e.g., *sp_cdhs, sp_eps*) that is displayed as a string floating near the standard port (either inside or outside the block border). A standard port can have

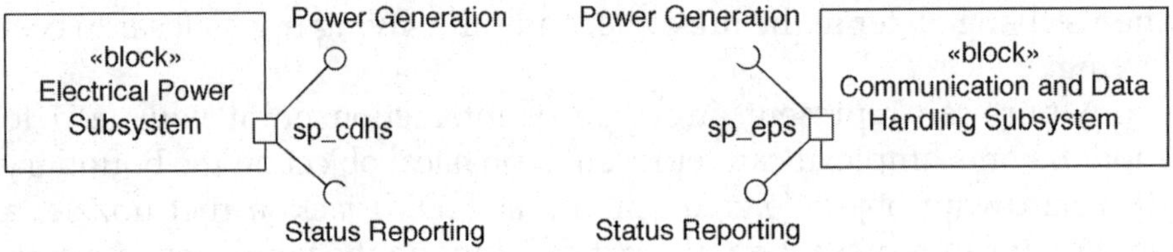

Figure 3.9 *Blocks with standard ports*

one or more types; the types are the **interfaces** you assign to it (e.g., *Power Generation, Status Reporting*).

An interface, like a block, is an element of definition—one that defines a set of **operations** and **receptions,** a behavioral contract that clients and providers will conform to. You can display an interface on a BDD as a rectangle with the keyword «interface» preceding the name; you can display its operations and receptions in the second and third compartments. Figure 3.10 displays the *Power Generation* and *Status Reporting* interfaces using this notation.

When you assign an interface to a standard port, you assign it either as a provided interface or as a required interface. A **provided** interface is displayed using the ball notation—the lollipop symbol attached to the standard port (shown in Figure 3.9). A block that provides an interface must implement *all* of the interface's operations and receptions. For example, Figure 3.9 conveys that the *Communication and Data Handling Subsystem* block provides the *Status Reporting* interface, and

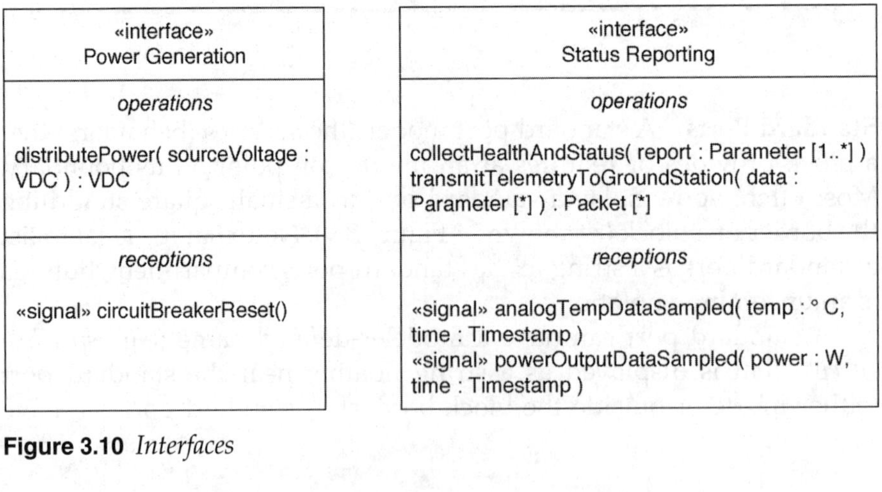

Figure 3.10 *Interfaces*

this means that it implements (can perform) the two operations and the two receptions in that interface.

A **required** interface is displayed using the socket notation—the stick with a semicircle attached to the standard port (shown in Figure 3.9). A block that requires an interface may invoke one or more—but not necessarily all—of its operations or receptions at some point during system operation. For example, Figure 3.9 conveys that the *Electrical Power Subsystem* block requires the *Status Reporting* interface, and this means that it may invoke any (or all) of the four operations and receptions in that interface.

Modeling with standard ports and interfaces is a way to decouple clients and providers, enabling you to design to abstractions rather than specific implementations. This **extensibility** lets you add new providers of interfaces at any time without impacting the existing clients of those interfaces.

Flow Ports A **flow port** models the types of matter, energy, or data that can flow in or out of a block at an interaction point on its boundary. As with a standard port, you most often display a flow port as a small square straddling the border of a block (as shown in Figure 3.11). Unlike a standard port, however, a flow port has a symbol shown inside the small square (more on that soon). It's legal to list a flow port as a string in a compartment—one named "flow ports"—but again, this is an uncommon notation.

A flow port can have a modeler-defined name (e.g., *dataOut, dataIn*); it can also have a type (e.g., *Housekeeping Data*). The name and type are displayed as a string floating near the flow port, separated by a colon in the format *name : type.* The type that you specify for a flow port and the symbol that appears inside the square depend on the kind of flow port you're modeling. SysML offers two kinds of flow ports: nonatomic flow ports and atomic flow ports.

Figure 3.11 shows examples of nonatomic flow ports. You add a **nonatomic** flow port (symbolized as < >) to a block when you need to

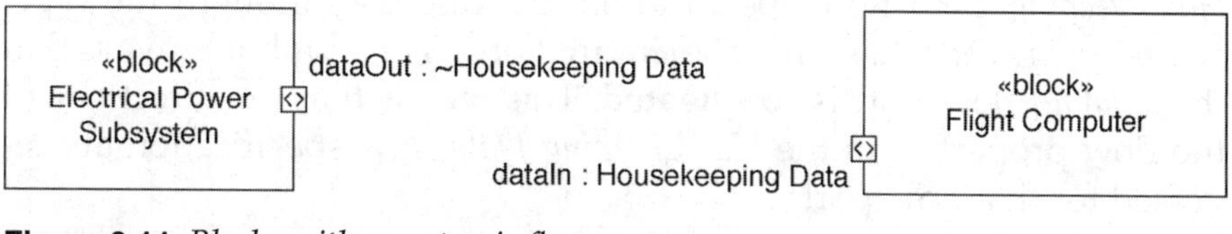

Figure 3.11 *Blocks with nonatomic flow ports*

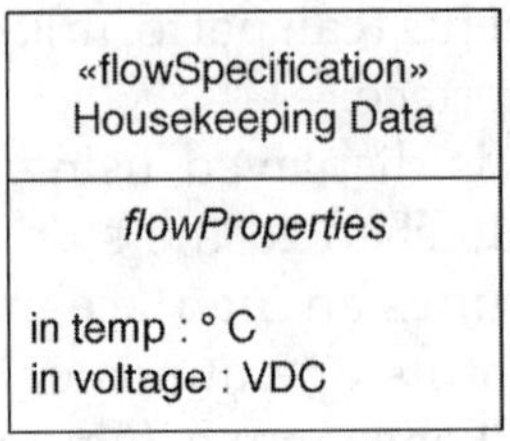

Figure 3.12 *A flow specification*

model *multiple* types of items that could flow in or out via that port. The type of a nonatomic flow port must be the name of a flow specification that you've created somewhere in the system model.

Like a block, a **flow specification** is an element of definition—one that defines a set of **flow properties** that can flow in or out of a nonatomic flow port. You can display a flow specification on a BDD as a rectangle with the stereotype «flowSpecification» preceding the name; you can display its flow properties in a compartment named "flowProperties." Figure 3.12 displays the *Housekeeping Data* flow specification using this notation.

A flow property represents a specific item that can flow in or out of a block via a flow port. Each flow property has a direction, a name, and a type, which are displayed as a string in the following format:

```
<direction> <name> : <type>
```

The direction can be *in, out,* or *inout*. The name is modeler defined. The type must be the name of a value type, block, or signal that you've created somewhere in your model hierarchy.

Figure 3.11 shows that the *Flight Computer* block owns a nonatomic flow port named *dataIn,* which is typed by the *Housekeeping Data* flow specification. This model conveys that temperature and voltage values can flow into an instance of *Flight Computer* at some point during system operation.

Figure 3.11 also shows that the *Electrical Power Subsystem* block owns a nonatomic flow port named *dataOut,* which also is typed by the *Housekeeping Data* flow specification. In this case, though, the type, *Housekeeping Data,* has a tilde (~) in front of it. This symbol conveys that the *dataOut* flow port is **conjugated**. This means that the directions of the flow properties in the *Housekeeping Data* flow specification are reversed for that flow port.

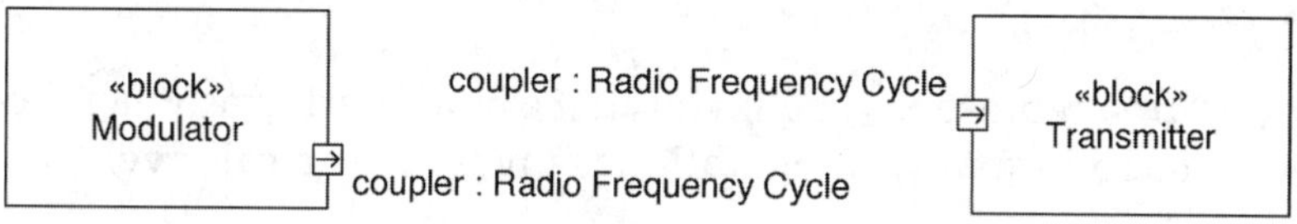

Figure 3.13 *Blocks with atomic flow ports*

The other kind of flow port is an atomic flow port. Figure 3.13 shows examples of this kind. You add an **atomic** flow port to a block when you need to model a *single* type of item that could flow in or out via that port. The symbol inside the small square is an arrow that conveys the direction of flow. The type of an atomic flow port must be the name of a value type, block, or signal that you've created somewhere in your model hierarchy.

Figure 3.13 shows that the *Modulator* block and the *Transmitter* block have an atomic flow port named *coupler,* which is typed by the same value type, *Radio Frequency Cycle.* These ports differ only in their direction of flow. This model conveys that a radio frequency signal can flow from a modulator to a transmitter via a coupler—an interaction point at their respective boundaries.

3.4.2 Behavioral Features

All the features I discuss in the preceding section are structural features. On most modeling projects, however, it's not sufficient to specify only the parts, references, constraints, value properties, and ports of a block. They're important, but they convey only one aspect of the design. An equally important aspect is the set of **behaviors** that a block can perform. You convey this aspect of the design by adding behavioral features to a block.

SysML offers two kinds of behavioral features: operations and receptions. I discuss these briefly in the context of interfaces earlier in Section 3.4.1.5. However, they're not limited to interfaces; you can also add operations and receptions to blocks. The decision to add a behavioral feature to a block directly or to an interface (that a block provides or requires) is a matter of your chosen modeling methodology and design principles. SysML does not dictate either course of action, and the format for displaying operations and receptions is the same in either case.

Now let's take a look in detail at each kind of behavioral feature.

3.4.2.1 *Operations*

An **operation** represents a behavior that a block performs when a client calls it. Stated formally, an operation is invoked by a **call event**.

Note

The term *call event* becomes more meaningful when I discuss events in detail in the context of behaviors in Chapter 6, "Activity Diagrams," Chapter 7, "Sequence Diagrams," and Chapter 8, "State Machine Diagrams." I introduce the term now to establish its connection to the concept of operations. As you study SysML incrementally, remember that diagrams are merely views of an underlying model; what you see on BDDs is related to what you see on other kinds of diagrams, including activity diagrams, sequence diagrams, and state machine diagrams.

Most often, an operation represents a **synchronous** behavior. This means that the caller waits for the behavior to complete before continuing with its own execution. However, SysML doesn't require this; you're free to represent any behavior as an operation—even when the caller doesn't wait for it to complete.

You display an operation on a BDD as a string in the operations compartment of a block (as shown in Figure 3.14). That string has the following format:

```
<operation name> ( <parameter list> ) : <return type>
  [<multiplicity>]
```

The operation name is modeler defined. The parameter list is a comma-separated list of zero or more parameters. (The format for each parameter is shown shortly.) The return type (if any) must be the name of a value type or block that you've created somewhere in your system model. The multiplicity is a constraint on the number of instances of the return type that the operation can return to the caller when it completes.

The parameters in the parameter list represent the inputs or outputs of the operation. Each parameter in the list is displayed with the following format:

```
<direction> <parameter name> : <type> [<multiplicity>] =
  <default value>
```

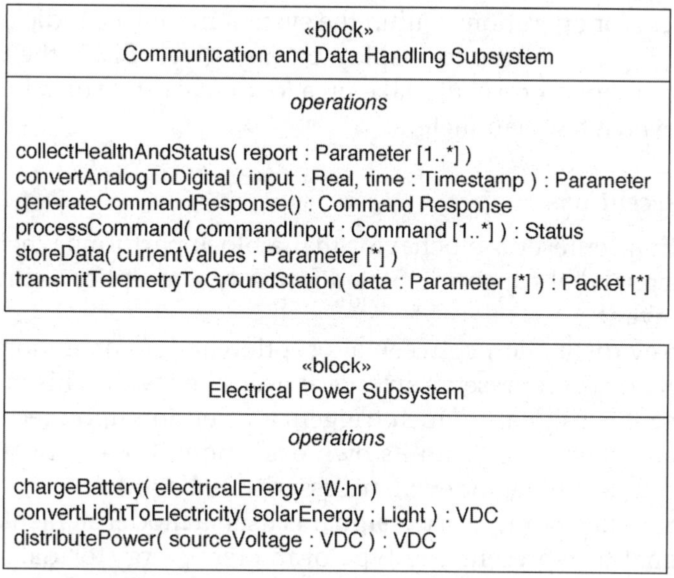

Figure 3.14 *Blocks with operations*

The direction can be *in, out,* or *inout.* The parameter name is modeler defined. The type must be the name of a value type or block that exists somewhere in your model. The multiplicity is a constraint on the number of instances of the type that the parameter can represent. The default value is the value assigned to the parameter if no value is specified as an argument when the operation is called.

Figure 3.14 shows that the *Electrical Power Subsystem* block and the *Communication and Data Handling Subsystem* block own several operations each—operations that represent behaviors that instances of these blocks can perform if called upon during system operation. An example of an operation is *processCommand*. This model conveys that a client can call the communication and data handling subsystem to perform this operation. When it does, the client can pass one or more commands as an input to the operation. And when the operation completes, it will return a status value to the caller.

A bit of advice: It's good practice to always use a verb phrase (such as *processCommand*) to name an operation; an operation represents a behavior, after all. Also, don't go overboard with the parameter lists; simply adding operations to blocks (without specifying parameters) is often a sufficient degree of model fidelity. If your team needs to specify parameters for operations within the system model, be judicious about which ones you choose to display on any given BDD; the complete string for an operation could take up a lot of real estate on a BDD if you display even a few parameters.

3.4.2.2 Receptions

A **reception** represents a behavior that a block performs when a client sends a signal that triggers it. Stated formally, a reception is invoked by a **signal event**.

The key distinction between a reception and an operation is that a reception always represents an *asynchronous* behavior. This means that a client sends a signal—which triggers a reception upon receipt—and immediately continues with its own execution; it doesn't wait for the reception to complete (or even, necessarily, to begin).

Another key point is that a **signal** is itself a model element. You can use a signal to represent any type of matter, energy, or data that one part of a system sends to another part—generally for the purpose of triggering a behavior on the receiving end. Like a block, a signal can own properties. Most often, those properties represent data that the signal carries from a client to a target. And when the signal arrives at the target and triggers a reception, the signal's properties become inputs to that reception.

Figure 3.15 displays a signal named *AnalogTempDataSampled*. This signal owns two properties: *temp* (of type ° C) and *time* (of type *Timestamp*). When a client generates an instance of this signal during system operation, it can supply values for the two properties. The client can send the signal instance to a target that's receptive to it (e.g., the *Communication and Data Handling Subsystem* block shown in Figure 3.16).

A structure is an eligible target for a signal if it owns a reception that has the same name as the signal. Additionally, the reception must

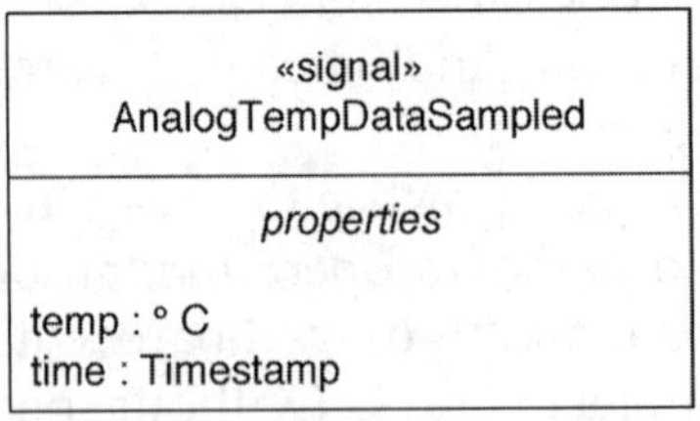

Figure 3.15 *A signal*

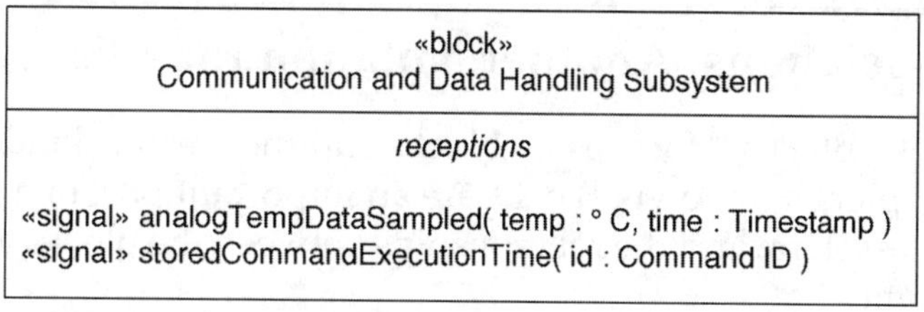

Figure 3.16 *A block with receptions*

have a parameter with a compatible type for each property of the signal. The *Communication and Data Handling Subsystem* block meets these criteria. When an instance of this block in an operational system receives an instance of the *AnalogTempDataSampled* signal, the reception behavior gets invoked, and the values held in the signal's two properties become inputs to that behavior.

When you display a reception in the receptions compartment of a block, the string has the following format:

```
«signal» <reception name> ( <parameter list> )
```

The keyword «signal» must always precede the reception name. As mentioned earlier, the reception name must match the name of the signal in your model that triggers it. You can display as many parameters as necessary in the parameter list. Each parameter in the list is displayed with the following format:

```
<parameter name> : <type> [<multiplicity>] = <default value>
```

The parameter name is modeler defined. The type must be the name of a value type or block that exists somewhere in your model. The multiplicity is a constraint on the number of instances of the type that the parameter can represent. The default value is the value assigned to the parameter if no value is provided in the corresponding property of the signal.

Unlike operations, receptions cannot have return types. Receptions are asynchronous; the client that sent the signal isn't waiting for a reply. For the same reason, the parameters of a reception can only be inputs and never outputs.

3.5 Associations: Another Notation for a Property

Section 3.4, "Blocks," focuses on blocks and the various kinds of properties that blocks can own. Blocks are an important part of a structural model of a system, and the relationships between the blocks are at least as important.

There are three main kinds of relationships that can exist between blocks: associations, generalizations, and dependencies. I discuss generalizations and dependencies in detail in Section 3.6, "Generalizations," and Section 3.7, "Dependencies." This section is devoted to associations.

In discussing reference properties and part properties in Section 3.4.1, "Structural Features," I implicitly address the idea of **associations** between blocks. To reiterate the key points: A reference property represents a structure that's external to a block—a structure that the block needs to be connected to for some purpose. A part property instead represents a structure that's internal to a block—in other words, a structure that the block is composed of.

Reference properties and part properties correspond to two kinds of associations that you often create between blocks and display on BDDs: reference associations and composite associations, respectively. Associations are simply an alternative notation to convey these kinds of structural relationships within a system.

Let's take a look in detail at the two kinds of associations.

3.5.1 Reference Associations

A **reference association** between two blocks means that a connection can exist between instances of those blocks in an operational system. And those instances can access each other for some purpose across the connection.

The notation for a reference association on a BDD is a solid line between two blocks. An open arrowhead on exactly one end conveys unidirectional access; the absence of arrowheads on either end conveys bidirectional access.

The upper BDD in Figure 3.17 displays a reference association between the *Electrical Power Subsystem* block and the *Flight Computer* block. Associations can have several labels. You can optionally display an association name floating near the middle of the line, and you can optionally display a role name and multiplicity on either end of the line. The association name is a modeler-defined string that describes

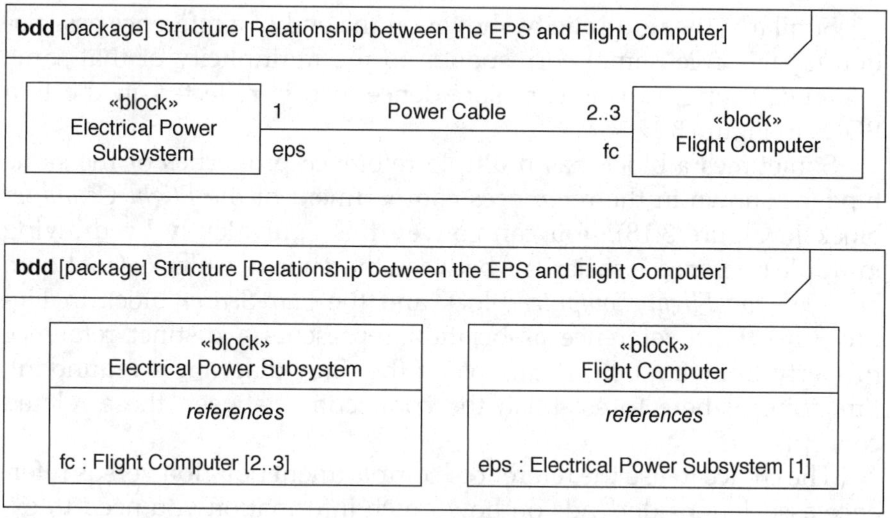

Figure 3.17 *Reference associations and reference properties*

the type of connection that can exist between instances of the two blocks. In Figure 3.17, for example, the name of the reference association shown is *Power Cable*—a name that describes the type of connection that could exist between an electrical power subsystem and a flight computer in a correctly assembled satellite.

Note

I use the phrase "type of connection" deliberately. An association is an element of definition; it can serve as the type for one or more connectors. A **connector** is an element of usage that appears on internal block diagrams (IBDs) (more on this in Chapter 4).

The role name shown on the end of a reference association corresponds to the name of a reference property—one that's owned by the block at the *opposite* end and whose type is the block that it's *next* to. In the upper BDD in Figure 3.17, for example, the role name *eps* represents a reference property that belongs to the *Flight Computer* block and whose type is the *Electrical Power Subsystem* block. The role name *fc* represents a reference property that belongs to the *Electrical Power Subsystem* block and whose type is the *Flight Computer* block. The lower BDD in Figure 3.17 displays an equivalent view of the same model using the references compartment notation instead of reference associations.

Similarly, the multiplicity shown on the end of a reference association (near a role name) corresponds to the multiplicity of that same reference property. This correspondence also is reflected in the two BDDs in Figure 3.17.

Sometimes a block has multiple reference properties of the same type (as shown in the references compartment of the *Flight Computer* block in Figure 3.18). You can convey this equivalently by drawing multiple reference associations between the same two blocks (as shown between the *Flight Computer* block and the *Star Sensor* block in Figure 3.18). Each reference association represents a distinct reference property. Showing both notations on the same diagram is redundant; I'm doing it here to establish the connection between these related concepts.

The choice to use the references compartment notation versus reference associations depends on how much information you need to expose on the BDD. In Figure 3.18, for example, the reference association notation lets me expose the value properties of the *Star Sensor* block; in contrast, the references compartment notation hides all features of the *Star Sensor* block.

Another factor in your decision is the need to specify a type for a connector on an IBD. If you intend to do this, then you need to create a reference association between two blocks and give it a name. The compartment notation would not meet your needs in this case.

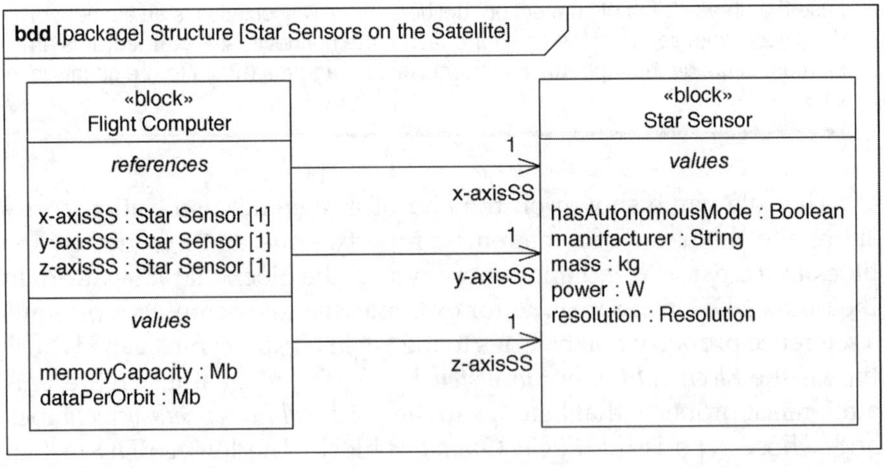

Figure 3.18 *Using reference associations to specify multiple reference properties of the same type*

3.5.2 Composite Associations

A **composite association** between two blocks conveys structural decomposition. An instance of the block at the composite end is made up of some number of instances of the block at the part end.

The notation for a composite association on a BDD is a solid line between two blocks with a solid diamond on the composite end. An open arrowhead on the part end of the line conveys unidirectional access from the composite to its part; the absence of an arrowhead conveys bidirectional access (i.e., the part will have a reference to the composite).

Figure 3.19 displays four examples of composite associations from the *DellSat-77 Satellite* block to the subsystem blocks. (It's permissible and common practice to overlap the solid diamonds on the composite end.) This BDD conveys that a correctly manufactured and assembled DellSat-77 satellite will be composed of one electrical power subsystem, one attitude and orbit control subsystem, one environmental

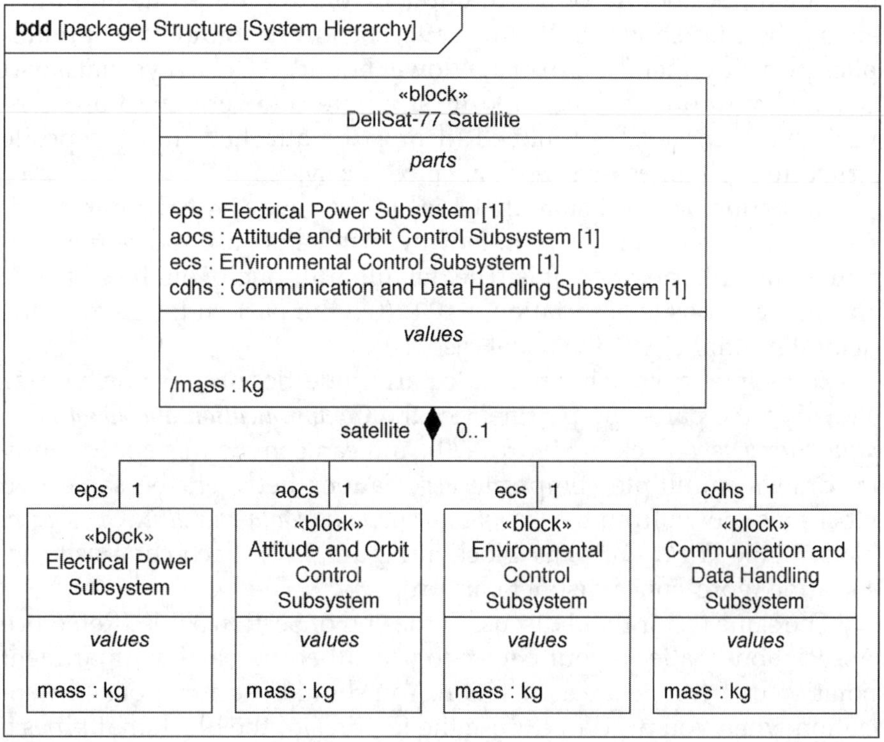

Figure 3.19 *Composite associations and part properties*

control subsystem, and one communication and data handling subsystem. The possible numbers of instances are conveyed by the multiplicities on the part ends of the four composite associations.

The role name shown on the part end of a composite association corresponds to the name of a part property—one that's owned by the block at the composite end and whose type is the block at the part end. In Figure 3.19, for example, the role name *aocs* represents a part property that's owned by the *DellSat-77 Satellite* block and whose type is the *Attitude and Orbit Control Subsystem* block. This correspondence is equivalently reflected in the parts compartment of the *DellSat-77 Satellite* block. It's redundant to show both the parts compartment notation and composite associations on the same diagram; I'm doing it here to reinforce the connection between these concepts.

The multiplicity on the part end of a composite association is not restricted; a composite structure can be made up of an arbitrary number of instances of parts—however many a system requires.

However, the multiplicity on the composite end is restricted. A part—by definition—can belong to only one composite at a time. Therefore, the upper bound of the multiplicity on the composite end must always be 1 (as shown in Figure 3.19). The lower bound of that multiplicity can be either 0 (zero) or 1. A lower bound of 0 conveys that a part can be removed from its composite structure; a lower bound of 1 conveys that it cannot be removed (it must be attached to a composite structure at all times in a valid instance of a system).

In Section 3.4.1.1, I state that 1 is almost always the default multiplicity for elements in SysML. However, there is an important exception to this rule, and here it is: The default multiplicity on the *composite* end of a composite association is 0..1. (On the part end, however, the default multiplicity is the usual case, 1.)

Sometimes a block has multiple part properties of the same type (as shown in the parts compartment of the *Communication and Data Handling Subsystem* block in Figure 3.20). You can convey this equivalently by drawing multiple composite associations between the same two blocks (as shown from the *Communication and Data Handling Subsystem* block to the *Flight Computer* block in Figure 3.20). Each composite association represents a distinct part property.

The same factor that I discuss at the end of Section 3.5.1, "Reference Associations," affects your choice to use either the parts compartment notation or a composite association. You should use a composite association when you need to expose the features of the block that types a part; you should use compartment notation instead when those features are not the focus of the diagram.

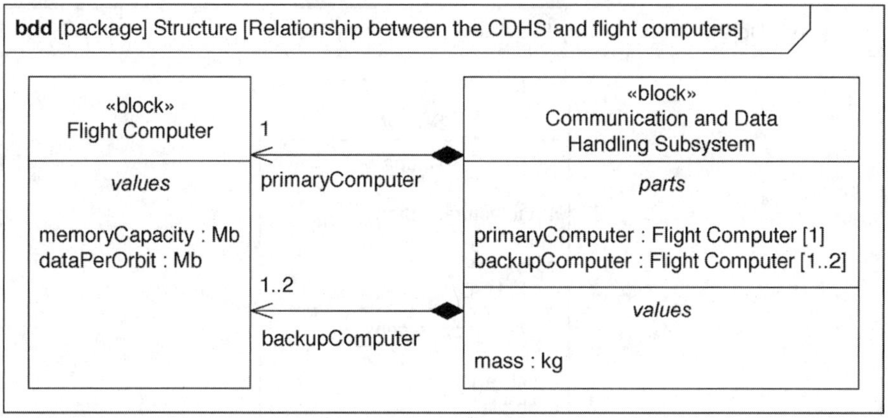

Figure 3.20 *Using composite associations to specify multiple part properties of the same type*

3.6 Generalizations

A **generalization** is another kind of relationship you typically display on BDDs. This relationship conveys **inheritance** between two elements: a more *generalized* element, called the **supertype**, and a more *specialized* element, the **subtype**. You use generalizations to create classification trees (type hierarchies) in your system model.

The notation for a generalization is a solid line with a hollow, triangular arrowhead on the end of the supertype. This relationship is read in English as "is a type of" going from the subtype to the supertype. For example, the BDD in Figure 3.21 shows a generalization from the *Gyroscope* block to the *Sensor* block (among others). This relationship conveys that a gyroscope *is a type of* sensor.

When a supertype has more than one subtype shown on the same BDD, modelers often overlap the hollow, triangular arrowheads on the supertype end to conserve space on the diagram (as shown in Figure 3.21). Purists will tell you that overlapping the arrowheads actually conveys a special grouping of subtypes called a **generalization set**. This is a slightly more advanced feature of the language that you may find useful later. For now, feel free to overlap the arrowheads purely to enhance the readability of your diagrams.

One key point is that generalizations are transitive. The model displayed in Figure 3.21 shows that a star mapper is a type of star sensor, and a star sensor is a type of sensor. Therefore, a star mapper is a type of sensor. Type hierarchies in your model can be arbitrarily deep.

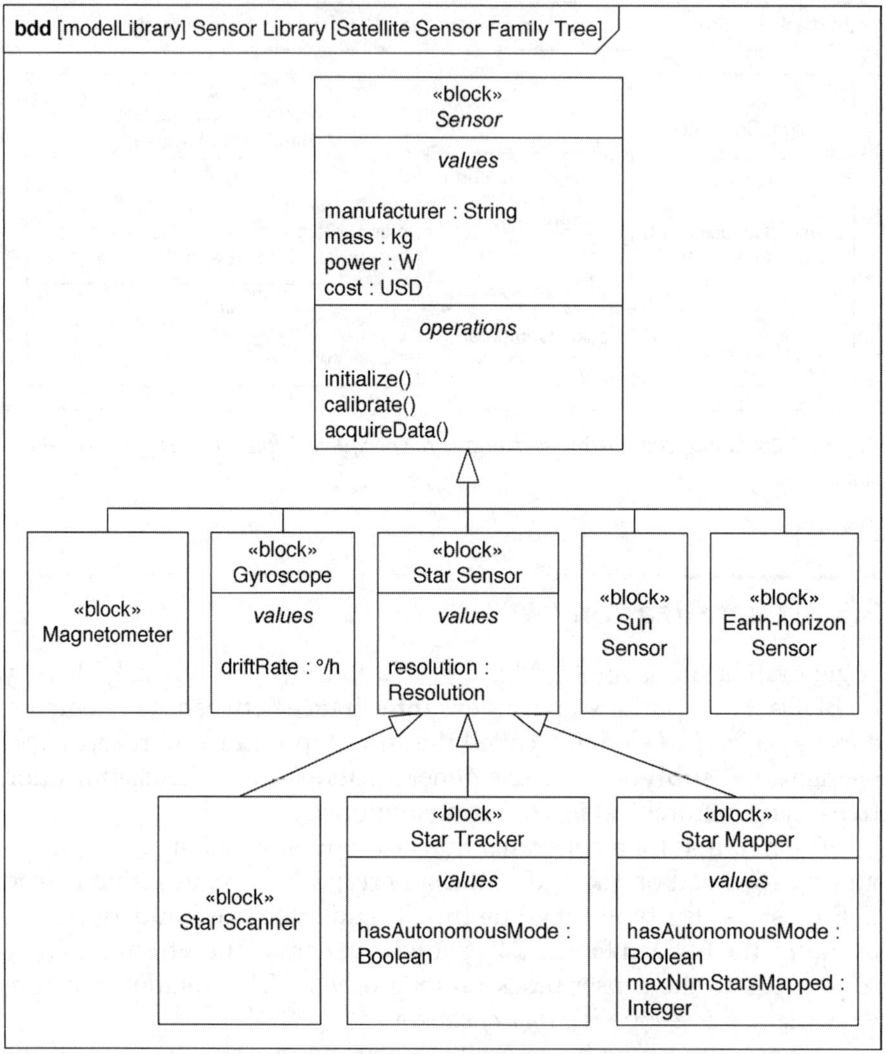

Figure 3.21 *Generalization relationships between blocks*

A generalization conveys that a subtype inherits *all* the features of its supertype: the structural features (properties) and the behavioral features (operations and receptions). In addition to the features it inherits, a subtype may have other features that its supertype doesn't have. For this reason, modelers often refer to a subtype as a **specialization** of its supertype.

For example, the *Star Sensor* block is a specialization of the *Sensor* block. It inherits the four value properties and three operations from the

Sensor block, and then it adds a fifth value property, *resolution*, that the *Sensor* block doesn't have. Similarly, the *Star Mapper* block inherits the five value properties and three operations from the *Star Sensor* block, and then it adds two new value properties (*hasAutonomousMode* and *maxNumStarsMapped*), which neither of its supertypes have.

You create generalizations to define **abstractions** in your system design. A supertype (such as *Sensor*) is an abstraction of its subtypes; it factors out those features that are common among the subtypes. Abstractions let you define a common feature (such as the *initialize* operation) in one place within the model—in the supertype—and that common feature propagates down the type hierarchy to all the subtypes. Then, if you later need to change that common feature, you simply go back to that one place in the model to make the change, and all subtypes in the model get updated instantly.

Abstraction is a powerful design principle; it conveys **substitutability**, meaning that a subtype will be accepted wherever its supertype is required. For example, Figure 3.22 shows that the *Flight Computer*

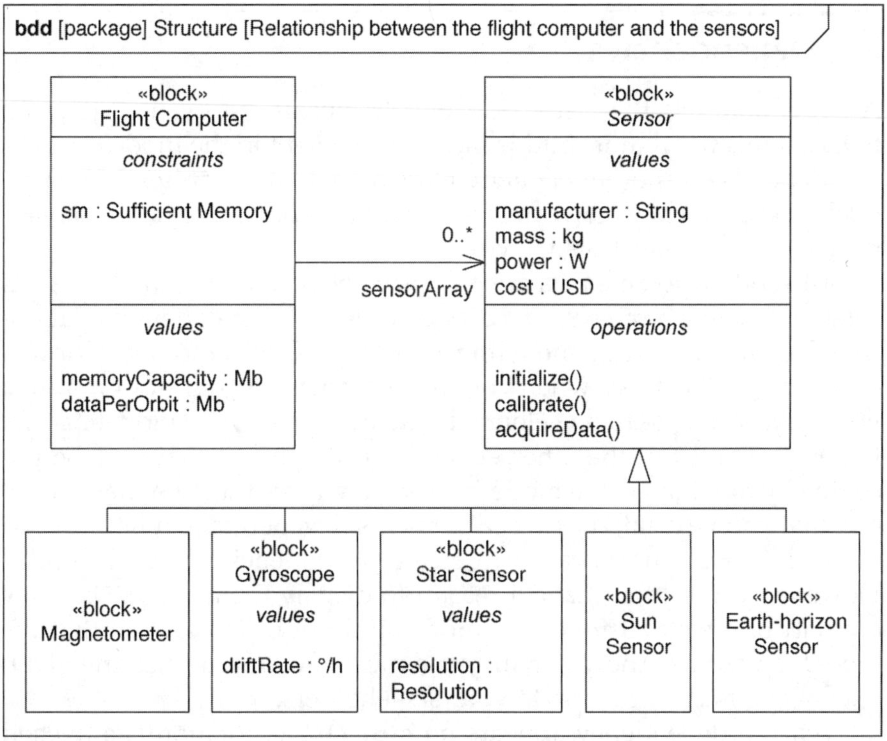

Figure 3.22 *Designing to an abstraction*

block has a reference property named *sensorArray* of type *Sensor*. This model conveys that a flight computer may need access to one or more of the features—structural or behavioral—that are common to all sensors. Therefore, any of the five subtypes of *Sensor* would be acceptable to a flight computer, because all of them inherit those common features from their supertype, *Sensor*.

This is an example of designing to an abstraction. This practice creates extensibility in your design. When the customers' requirements change later in the life cycle and you need to add a new type of sensor to the satellite design, you can simply define a new subtype of the *Sensor* block within the system model, and that addition will be transparent to all clients (such as *Flight Computer*) that reference the *Sensor* block. For all these reasons, building generalizations into your model can significantly reduce the time it takes to modify your system design as the life cycle progresses—and that capability directly translates into cost savings.

3.7 Dependencies

A **dependency** is the third kind of relationship you can display on BDDs. It means what it sounds like: One element in the model, the **client**, depends on another element in the model, the **supplier**. More precisely, a dependency conveys that when the supplier element changes, the client element *may* also have to change.

Most often, you create a dependency between two model elements solely to establish traceability between them. A dependency relationship lets you use your modeling tool to perform automated downstream impact analysis when you make changes to your design. When you make a change to one element, you can query your modeling tool to generate a list of the other elements in the model that may be impacted by the change; the modeling tool navigates the set of dependencies that you've created between elements to generate that list.

This is a practical reason to create dependencies in your model. However, you seldom have a reason to display them on BDDs. They are part of the structure of the model and not of the system that the model represents. And you will spend most of your time creating BDDs to convey system structure to your stakeholders.

When a dependency appears on a BDD, the notation is a dashed line with an open arrowhead, which is drawn *from* the client *to* the sup-

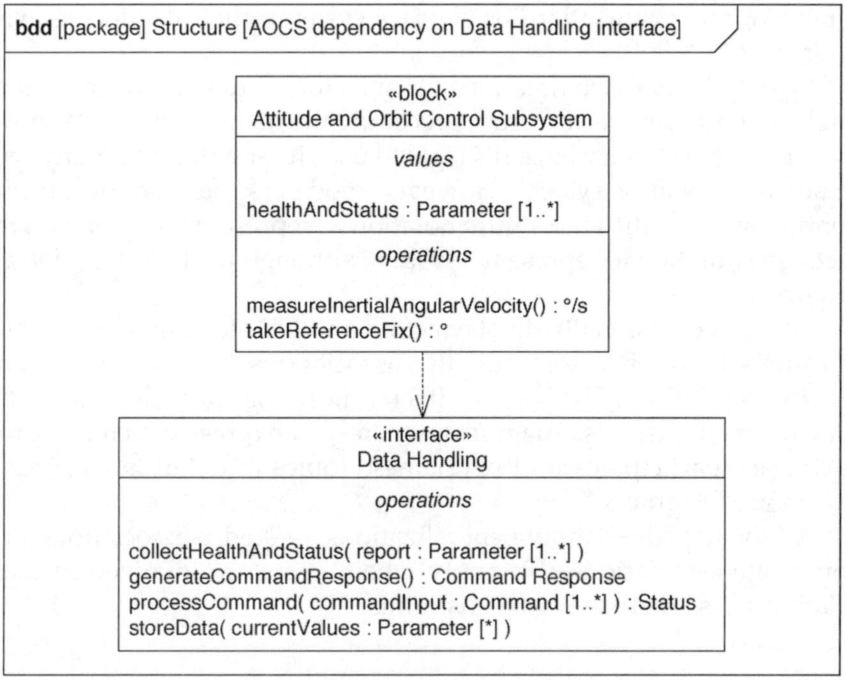

Figure 3.23 *A dependency relationship between two named elements*

plier. In Figure 3.23, for example, the *Attitude and Orbit Control Subsystem* block is the client, and the *Data Handling* interface is the supplier. This model conveys that the block depends on the interface; if the interface changes, the block may need to change, too.

Note that SysML defines specialized kinds of dependency relationships (e.g., package import, viewpoint conformance, and several kinds of requirements relationships). Although you rarely display dependencies on BDDs, you often display these specialized kinds of dependencies on package diagrams and requirements diagrams. I discuss these topics in detail in Chapter 10, "Package Diagrams," and Chapter 11, "Requirements Diagrams."

3.8 Actors

An **actor** represents someone or something that has an external interface with your system. The name of an actor conveys a **role** played by a person, an organization, or another system when it interacts with your system.

SysML defines two notations for an actor: a stick figure and a rectangle with the keyword «actor» preceding the name. Figure 3.24 shows examples of both notations. It's legal to use either notation for any type of actor—person or system. However, modelers often adopt the convention of using the stick figure notation to represent a person and the rectangle notation to represent a system, although the language doesn't require it.

You will occasionally display actors on BDDs to express the generalizations between actors and the associations between actors and blocks (as shown in Figure 3.24). It's far more common, though, to display actors on use case diagrams, where you express which use cases each actor participates in. I cover these topics in detail in Chapter 5, "Use Case Diagrams."

All the key ideas about generalizations, reference associations, and composite associations also apply when actors are involved in these relationships. There are two constraints:

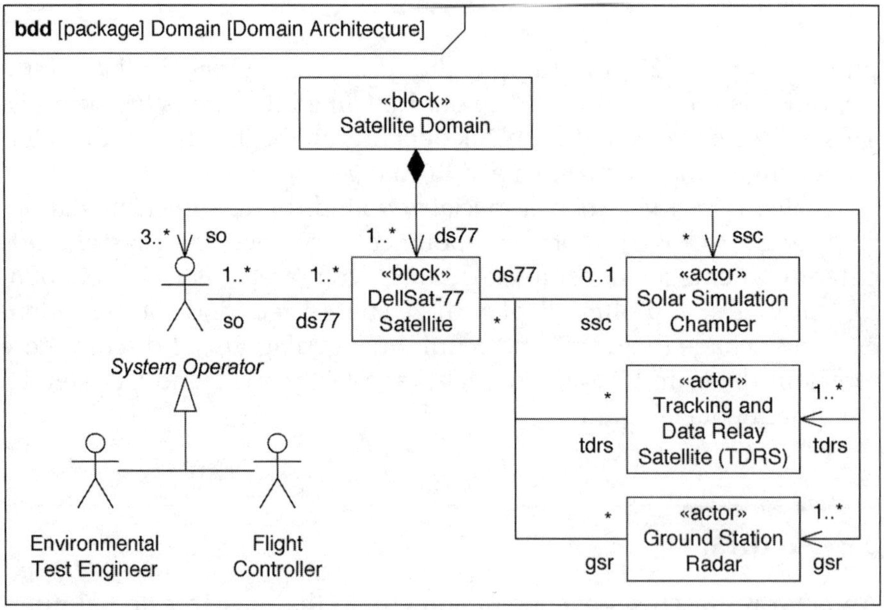

Figure 3.24 *Actors on a BDD*

- You cannot define a generalization between an actor and a block.
- An actor cannot have parts; that is, it cannot appear at the composite end of a composite association. (We always regard an actor as a "black box.")

3.9 Value Types

Like a block, a **value type** is an element of definition—one that generally defines a type of quantity. I say "generally" because there are two value types in SysML—*Boolean* and *String*—that arguably are not quantities.

You can use a value type in many places throughout your model. Most often, it appears as the type of a **value property**, which is a kind of structural feature of blocks. (Section 3.4.1.3, "Value Properties," has more details.) But that's not the only place where value types make an appearance; they're actually ubiquitous in system models. They can also appear as the types of the following:

- Atomic flow ports on blocks and actors
- Flow properties in flow specifications
- Constraint parameters in constraint blocks
- Item flows and item properties on connectors
- Return types of operations
- Parameters of operations and receptions
- Object nodes, pins, and activity parameters within activities

There are three kinds of value types—primitive, structured, and enumerated—that you typically define in your system model. A **primitive** value type has no internal structure (it doesn't own any value properties). Its notation is a rectangle with the stereotype «valueType» preceding the name.

SysML defines four primitive value types: *String, Boolean, Integer,* and *Real*. You can, of course, define your own primitive value types as specializations (subtypes) of these four. For example, Figure 3.25 shows three value types (°, *V*, and ° C) that are subtypes of *Real*.

As its name implies, a **structured** value type has an internal structure—generally two or more value properties. As with a primitive value type, the notation for a structured value type is a rectangle with

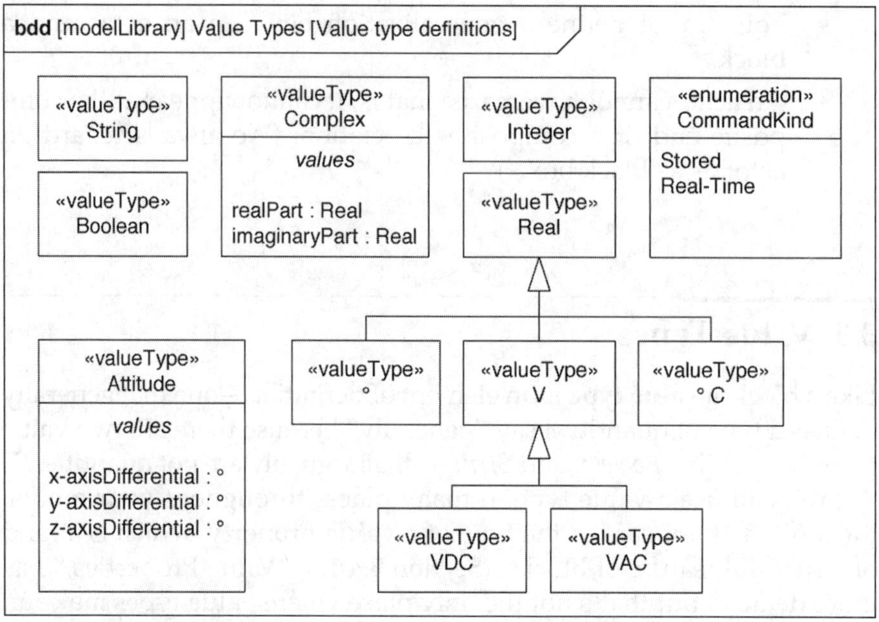

Figure 3.25 *Value types*

the stereotype «valueType» preceding the name. SysML defines one structured value type: *Complex*. Its structure consists of two value properties—*realPart* and *imaginaryPart*—that are both of type *Real*. One structured value type may, in turn, be the type of a value property within another structured value type. In this way, you can create arbitrarily complex systems of value types.

An **enumerated** value type—colloquially called an **enumeration**—simply defines a set of **literals** (legal values). If a parameter of an operation (or some other kind of element shown in the earlier bulleted list) is typed by an enumeration, then the value it holds at any moment must be one of the literals in that enumeration. The BDD in Figure 3.25 shows an enumeration named *CommandKind*, which defines two literals: *Stored* and *Real-Time*. I could use this enumeration, for example, to type an input parameter named *kind* in an operation named *buildCommand*. When a client calls this operation (within a running system), the only legal values it can pass are *Stored* and *Real-Time*.

I mentioned earlier that value types can be related to one another by using generalizations. A value type hierarchy can be arbitrarily deep, and generalizations—as you may recall—are transitive. For example, Figure 3.25 conveys that the value types *VDC* and *VAC* are (indirectly) subtypes of *Real*. The principle of substitutability applies here just as it does in the case of generalizations between blocks: Values of type *VDC* and *VAC* will be accepted wherever their supertypes (*V* and *Real*) are required. These supertypes are abstractions. And the principle of designing to an abstraction—and its consequent extensibility—also applies to this practice of creating a value type hierarchy. This is a widely used and powerful modeling practice.

3.10 Constraint Blocks

Like a block, a **constraint block** is an element of definition—one that defines a Boolean **constraint expression** (an expression that must evaluate to either *true* or *false*). Most often, the constraint expression you define in a constraint block is an equation or an inequality: a mathematical relationship that you use to constrain value properties of blocks. You would do this for two reasons:

- To specify assertions about valid system values in an operational system
- To perform engineering analyses during the design stage of the life cycle

The variables in a constraint expression are called **constraint parameters**. Generally, they represent quantities, and so they're typed most often by value types. For example, Figure 3.26 shows a constraint block named *Transfer Orbit Size,* which defines a constraint expression that contains three constraint parameters: *semimajorAxis, initialOrbitRadius,* and *finalOrbitRadius.* These three constraint parameters are typed by the value type *km*.

Constraint parameters receive their values from the value properties they're bound to—that is, the value properties that are being constrained. At any given moment, those values either satisfy the constraint expression, or they don't; the system is either operating nominally, or it isn't. Note, however, that a BDD by itself can't convey which constraint parameters and value properties are bound to one another. You would express this piece of information on a parametric diagram. (I discuss this in detail in Chapter 9.)

The notation for a constraint block on a BDD is a rectangle with the stereotype «constraint» preceding the name. The constraint expression always appears between curly brackets ({}) in the constraints compartment. The constraint parameters in the constraint expression are listed individually in the parameters compartment.

You sometimes build a more complex constraint block from a set of simpler constraint blocks. You would do this to create a more complex mathematical relationship from simpler equations and inequalities. The more complex constraint block can display its constituent parts as a list of **constraint properties** in the constraints compartment. Recall from Section 3.4.1.4 that a constraint property has a name and a type in the format *name : type*. The type, as mentioned earlier, must be the name of a constraint block.

For example, Figure 3.26 shows that the constraint block *Hohmann Transfer* is composed of two constraint properties—*ttof* and *tos*—which

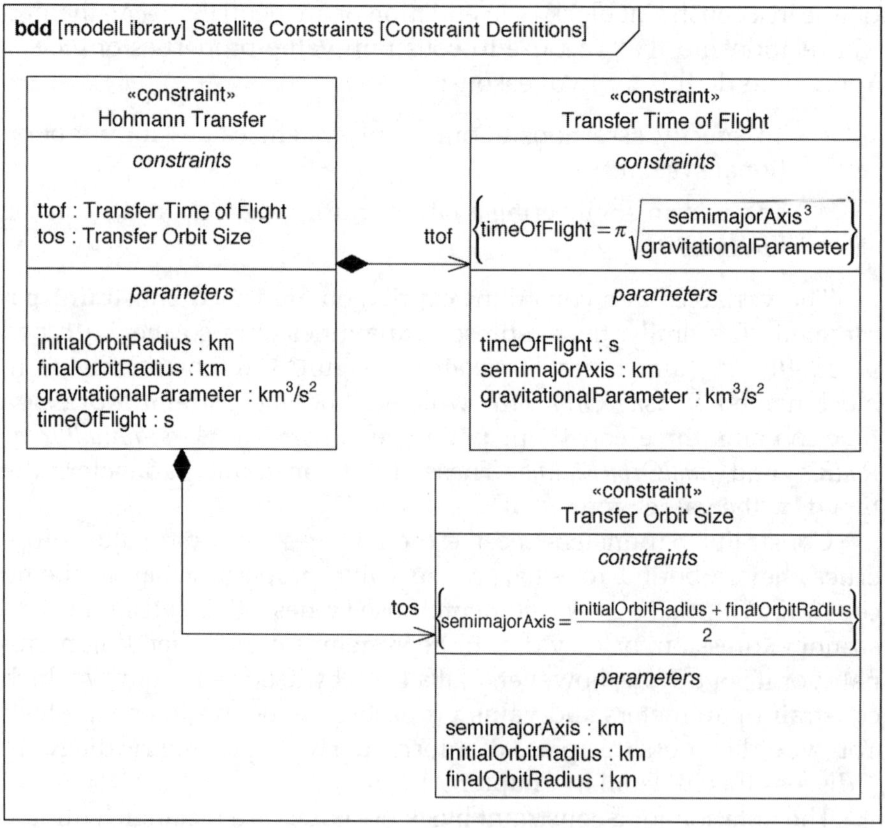

Figure 3.26 *Relationships between constraint blocks*

represent usages of the constraint blocks *Transfer Time of Flight* and *Transfer Orbit Size*, respectively. This model conveys that *Hohmann Transfer* defines a constraint expression that is a composite of two simpler constraint expressions—in effect, defining a more complex mathematical relationship.

Note, though, what this BDD doesn't (and can't) convey: *where* those two simpler constraint expressions are specifically connected to each other to create the composite constraint expression. A parametric diagram would convey this additional piece of information (more on this in Chapter 9).

As an alternative to the constraints compartment notation, you can use composite associations to convey that one constraint block is composed of other, simpler ones (as shown in Figure 3.26). Note that the role names shown on the part ends of the two composite associations correspond to the names of the constraint properties in the *Hohmann Transfer* constraint block. These are equivalent notations. You use composite associations when you need to expose the details of the simpler constraint blocks; in contrast, you use the constraints compartment notation to hide those details when they're not the focus of the diagram.

3.11 Comments

SysML has a lot of rules (and they all exist to serve the very useful purpose of giving your design unambiguous meaning from one reader to the next). However, you sometimes need to express information on a diagram in an unconstrained way as a block of text. You can do this with a comment.

A **comment** is, in fact, a model element. It consists of a single attribute: a string of text called the **body**. You can convey any information you need to in the body of a comment, and you can optionally attach a comment to other elements on a diagram to provide additional information about them. You can use comments on any of the nine kinds of SysML diagrams.

The notation for a comment is commonly referred to as a **note symbol**: a rectangle whose upper-right corner is bent. You use a dashed line to attach a comment to other elements (as shown at the bottom of the BDD in Figure 3.27). If you need to, you can attach a comment to several model elements simultaneously by using a separate dashed line for each one.

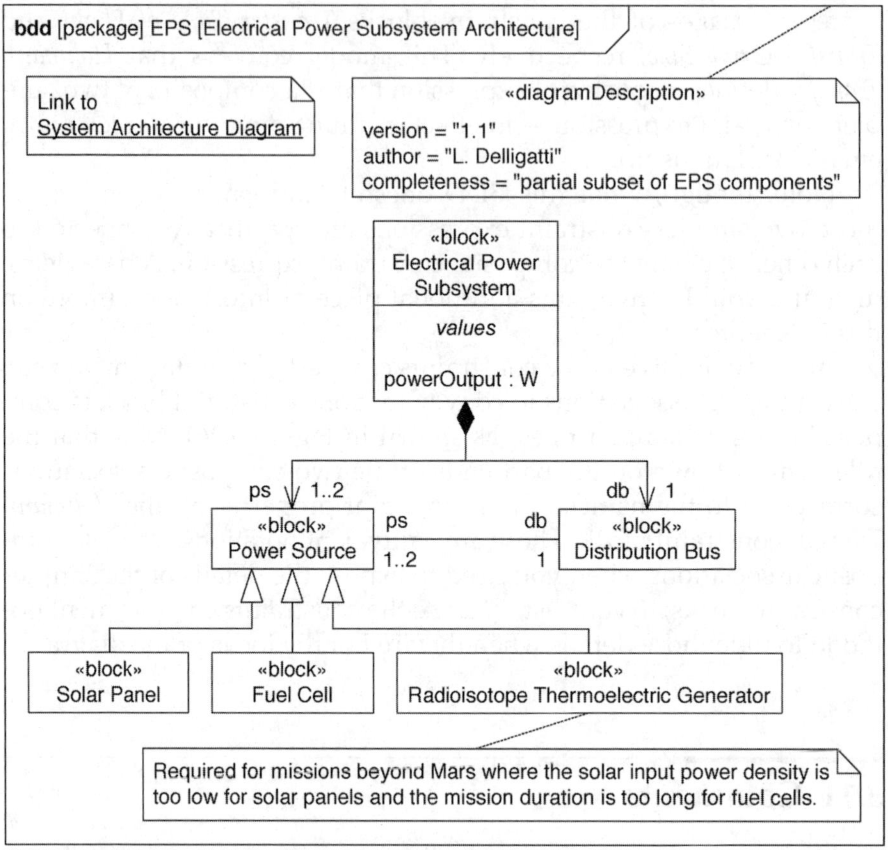

Figure 3.27 *Comments on a BDD*

Modelers sometimes put freestanding comments with hyperlinks on a diagram to enable readers to quickly navigate to a related diagram in the model (or to an external document). An example of this is shown in the upper-left corner of the BDD in Figure 3.27. To be clear, though, this capability is a function of the modeling tool you use; not all tools do this. And SysML itself says nothing about this capability.

SysML defines some specialized kinds of comments: rationale, problem, and diagram description. These appear as a note symbol with the respective stereotype preceding the body of the comment. Figure 3.27 shows an example of a diagram description comment in the upper-right corner of the BDD. Modelers often use rationale comments in conjunction with requirements relationships and allocations. I discuss these topics in detail in Chapters 11 and 12.

Summary

The BDD is the primary kind of diagram you create to communicate structural information about a system. A BDD enables you to express the types of structures that can exist internally within a system and externally in a system's environment. You can also use BDDs to express the types of services each structure provides and requires, the types of constraints each structure must conform to, and the types of values that can exist within an operational system.

Generalization relationships between elements let you define type hierarchies and design to abstractions. This is a powerful design technique—one that creates extensibility in your system design by decoupling the clients of services from any specific implementation of a provider of those services. As your stakeholders' requirements evolve over time, you can modify existing providers or add new ones with minimal impact on the rest of the system design.

Chapter 4

Internal Block Diagrams

The internal block diagram (IBD) has a close relationship to the BDD. You can display various elements on an IBD to express aspects of a system's structure that complement the aspects conveyed on BDDs. IBDs also have unique capabilities that make them an essential addition to your modeling toolbox.

4.1 Purpose

You create an IBD to specify the internal structure of a single block. Like a BDD, an IBD is a static (structural) view of the system or one of its parts. Unlike a BDD, an IBD does not display blocks; it displays *usages* of blocks—that is, the part properties and reference properties of the block that is named in the header of the IBD.

Recall from Chapter 3, "Block Definition Diagrams," that you can display part properties and reference properties on a BDD, too—either as strings in a block's compartments or as role names on the ends of associations. But an IBD enables you to convey additional information that you can't convey on a BDD: the connections among part properties and reference properties; the types of matter, energy, or data that flow across the connections; and the services that are provided and required across the connections.

An IBD conveys how the parts of a block must be assembled to create a valid instance of the block. It also shows how an instance of that block must be connected to external entities (reference properties) to create a valid instance of the system as a whole.

This is a powerful capability. It's important, however, to be aware of the limitations of SysML. SysML offers no means to model the geometry of your system. An IBD lets you model *which* parts must be connected to each other, but it doesn't let you model their shapes or the proper spacing between them. You would have to use a (non-SysML) computer-aided design (CAD) tool to accomplish that goal.

4.2 When Should You Create an IBD?

IBDs and BDDs provide complementary views of a block. A BDD lets you first define a block and its properties. Then you can use an IBD to display a valid **configuration** of that block—a specific set of connections among the block's properties. Because of this close relationship, you often create IBDs and BDDs in tandem for various stakeholders at different points in the system life cycle.

4.3 Blocks, Revisited

I provide a sample IBD in Section 4.5, "BDDs and IBDs—Complementary Views of a Block." Then in the remainder of the chapter, I discuss in detail the kinds of elements and notations that can appear on an IBD. Before I do that, though, let's revisit the subject of blocks.

Recall that blocks serve as *types* for the model elements that appear on IBDs. Blocks, however, cannot appear on IBDs; they appear on BDDs. Therefore, I begin by showing you a BDD that displays the subset of blocks used to create the IBD in Section 4.5. The BDD in Figure 4.1 is an excerpt of the larger BDD in Figure 3.1, which appeared at the beginning of Chapter 3. Note the names of the blocks and the relationships among them; you will see the names of these blocks again in the IBD in Figure 4.2.

The key point: A BDD and an IBD provide complementary views of a block.

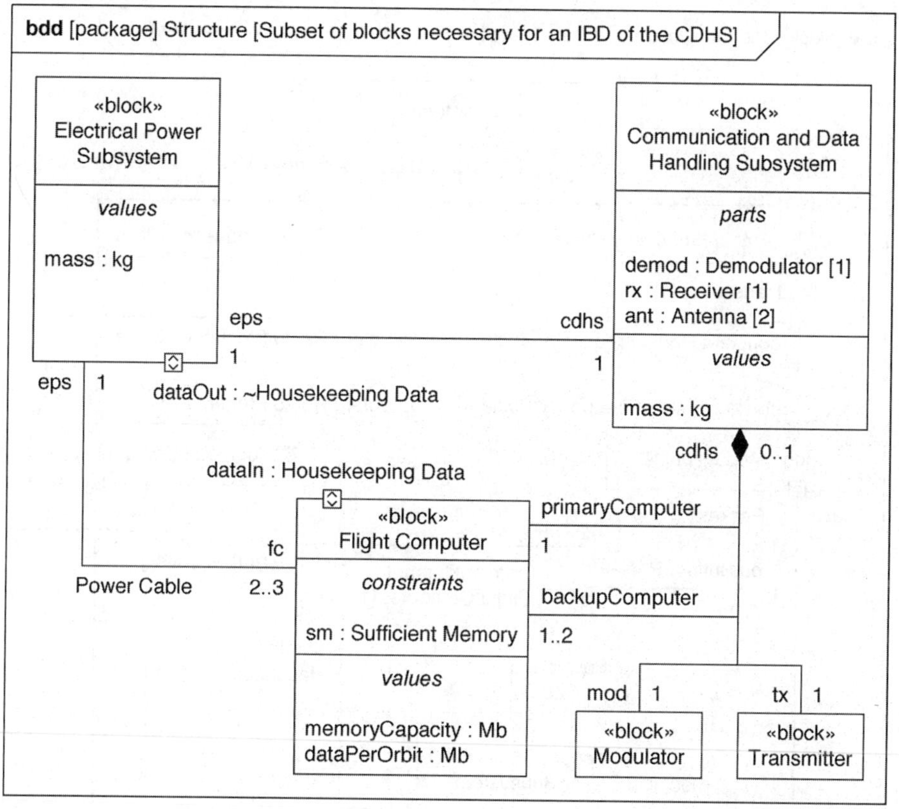

Figure 4.1 *The blocks necessary for an IBD of the* Communication and Data Handling Subsystem *block*

4.4 The IBD Frame

The diagram kind abbreviation for an internal block diagram is *ibd*. The only allowable *model element type* for an IBD is *block*. The frame of an IBD always represents a block that you've defined somewhere in your system model. Inside the frame, you can display that block's part properties and reference properties and the connectors that join them.

The name of the IBD in Figure 4.2 is "Flow-Oriented View." This IBD represents the *Communication and Data Handling Subsystem* block in the system model. That block, therefore, is the owner of the part properties and reference properties that appear on the diagram.

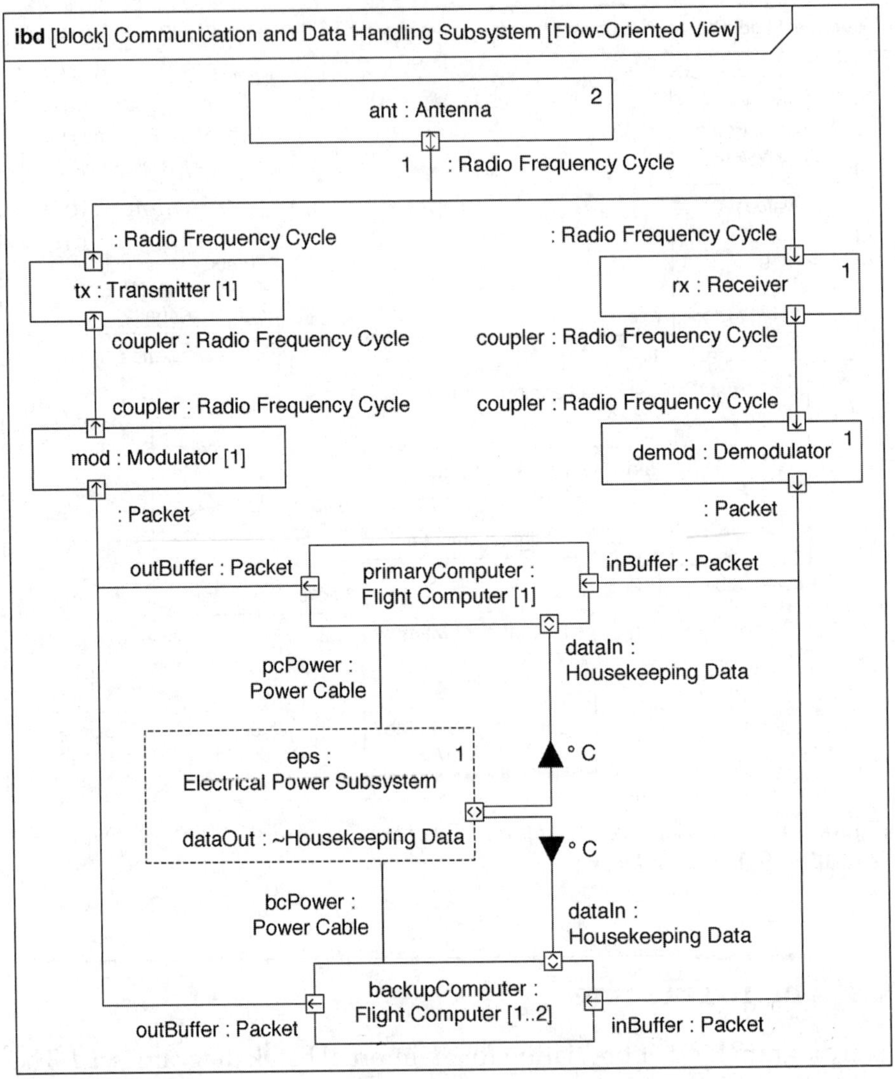

Figure 4.2 *A sample internal block diagram (IBD)*

4.5 BDDs and IBDs: Complementary Views of a Block

The BDD in Figure 4.1 conveys that the *Communication and Data Handling Subsystem* block has seven part properties: *demod*, *rx*, *ant*, *primaryComputer*, *backupComputer*, *mod*, and *tx*. It has one reference property: *eps*. These same eight properties appear on the IBD in Figure 4.2; the names, types, and multiplicities of these eight properties correspond between the two diagrams. Simply put, these diagrams present consistent and complementary views of the *Communication and Data Handling Subsystem* block.

The IBD in Figure 4.2 provides much of the same information as the BDD in Figure 4.1, but it provides some additional information that the BDD cannot: the specific connections among the internal part properties and their connections to the external reference property, *eps*. You use **connectors** between properties to convey that assembly.

In addition to the connectors between properties, IBDs can convey the items that flow among the properties and the services that properties invoke on one another across those connectors. You'll see each piece of this IBD in detail in the sections that follow.

4.6 Part Properties

A part property on an IBD has the same meaning as a part property in the parts compartment of a block on a BDD: It represents a structure that's internal to the block named in the IBD header—a structure that the block is composed of. The notation for a part property on an IBD is a rectangle with a solid border. The name string that appears inside the rectangle has the same format as the string that appears in the parts compartment of a block on a BDD:

```
<part name> : <type> [<multiplicity>]
```

You can optionally display a part property's multiplicity in the upper-right corner of the rectangle instead of at the end of the name string in square brackets. I show examples of both notations in Figure 4.2 for instructional purposes. However, I recommend using one notation consistently on the diagrams you create in your daily practice.

4.7 Reference Properties

A reference property on an IBD has the same meaning as a reference property in the references compartment of a block on a BDD: It represents a structure that's external to the block named in the IBD header—a structure that the block needs for some purpose, either to invoke behaviors or to exchange matter, energy, or data. The notation for a reference property on an IBD is a rectangle with a dashed border. The name string that appears inside the rectangle has the same format as the string that appears in the references compartment of a block on a BDD:

```
<reference name> : <type> [<multiplicity>]
```

As with a part property, you can optionally display a reference property's multiplicity in the upper-right corner of the rectangle.

4.8 Connectors

A connector between two properties on an IBD conveys that the two structures will have some way to access each other within a correctly assembled and operational system. You can optionally specify a name and type for a connector to convey additional information about the medium that connects those two structures. The format for that name string is as follows:

```
<connector name> : <type>
```

The connector name is optional and modeler defined. The type is also optional, but if you choose to specify one, it must be the name of an association you've created between two blocks somewhere in your system model. That **association** must relate the same two blocks that type the two properties at the ends of the connector.

The IBD in Figure 4.3 conveys that the flight computer (which is a *part* of the communication and data handling subsystem) is connected to the electrical power subsystem (which is a *reference* with respect to

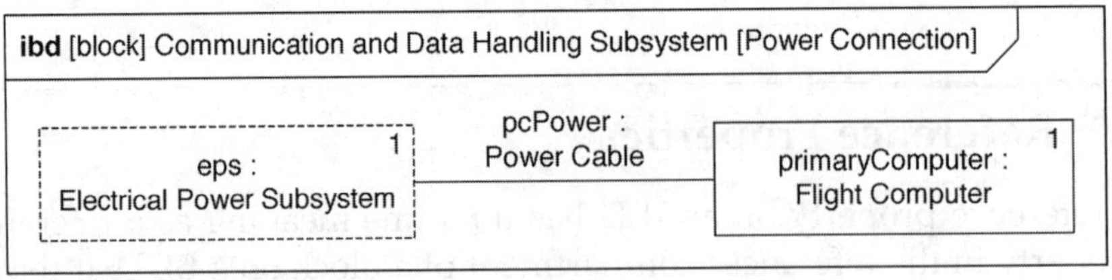

Figure 4.3 *A connector with a name and type*

the communication and data handling subsystem). The name of the connector is *pcPower*; its type is *Power Cable*. Specifying the name and type conveys additional information about the nature of the connection between these two structures. The type, *Power Cable*, corresponds to the name of the association between the *Electrical Power Subsystem* block and the *Flight Computer* block, as shown earlier on the BDD in Figure 4.1.

The two connected properties can be part properties, reference properties, or one of each. If the two connected properties have compatible ports—standard ports or flow ports—you can optionally attach the connector to those ports instead of to the properties directly. Doing so would convey that those properties are connected at specific points of interaction on their boundaries.

If you connect two properties via flow ports, you can convey the types of matter, energy, or data that can flow between the properties through those ports. In Figure 4.4, for example, the connector joins the *eps* reference property to the *primaryComputer* part property via nonatomic flow ports on their boundaries. These nonatomic flow ports are compatible, because they are typed by the same flow specification, *Housekeeping Data*, and one of the two ports, *dataOut*, is conjugated (as conveyed by the tilde [~] preceding its type). Recall from Chapter 3 that *conjugated* means the directions of the flow properties in the flow specification are reversed for that port.

If you connect two properties via standard ports, you can convey the services that each one provides and requires of the other at those ports. In Figure 4.5, for example, the connector joins the *eps* part property to the *cdhs* part property via standard ports on their boundaries. This IBD conveys that the electrical power subsystem provides the *Power Generation* interface and requires the *Status Reporting* interface. Reciprocally,

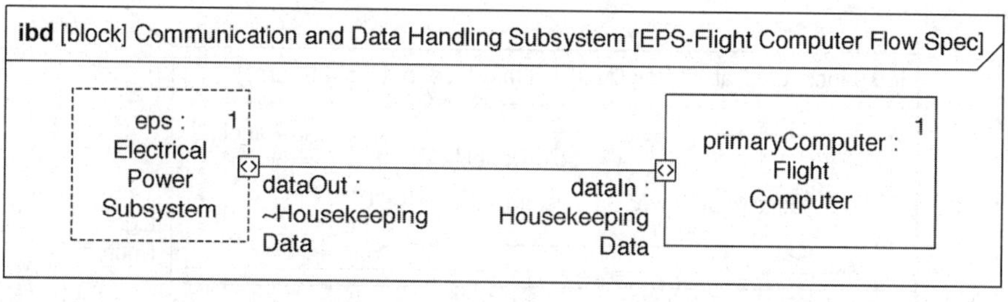

Figure 4.4 *A connector joining two properties via flow ports*

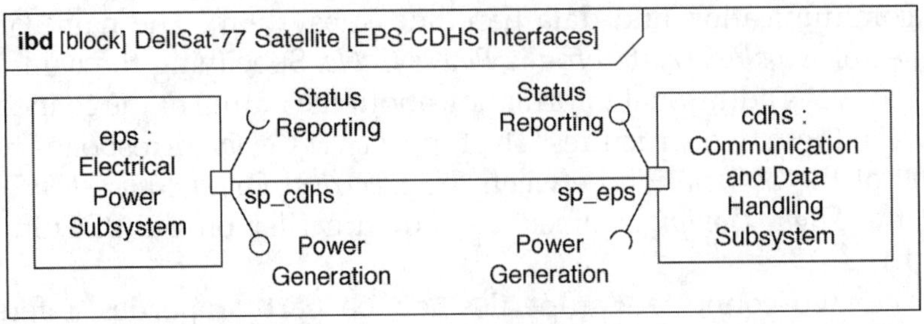

Figure 4.5 *A connector joining two properties via standard ports*

the communication and data handling subsystem provides the *Status Reporting* interface and requires the *Power Generation* interface. These standard ports are compatible, enabling these structures to exchange services across this connector during system operation.

You can also display ports on the frame of an IBD. Such ports represent points of interaction on the boundary of the block that the IBD represents—the one named in the diagram header. To convey that an internal part of a composite block is connected to the composite via some point of interaction at its boundary, you can connect a port on the frame to a port on a part property. This design conveys that an instance of the composite structure can pass requests for behaviors and item flows either from external clients to that internal part or from that internal part to external providers.

The IBD in Figure 4.6 shows that the *eps* part property (of the *DellSat-77 Satellite* block) is connected to the boundary of that block via the *solarPanel* standard ports. This diagram also shows that these standard ports have a required interface, *Light Source*. This model conveys that the satellite's electrical power subsystem requires a light source,

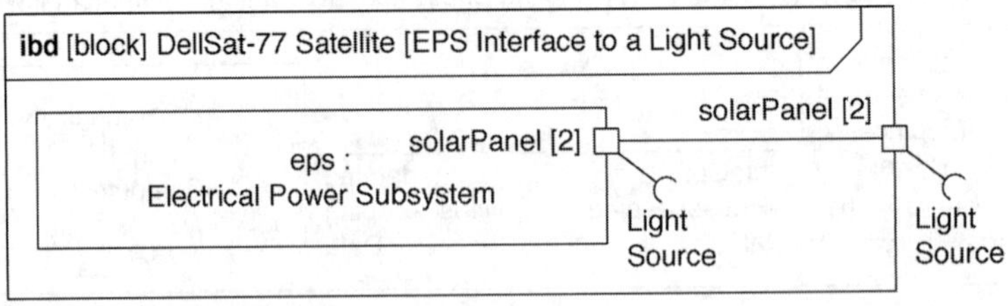

Figure 4.6 *An IBD with a port on the frame*

which it will access from the satellite's external environment via two solar panels on the satellite's boundary.

4.9 Item Flows

An **item flow** represents a type of matter, energy, or data that flows between two structures within a system. The notation for an item flow on an IBD is a filled-in triangular arrowhead on a connector that joins two flow ports (see Figure 4.7). The type that the item flow represents appears in a label near the arrowhead on the connector; the label must contain the name of a block, value type, or signal that exists somewhere in the system model.

The type that an item flow represents must be compatible with the types of the flow ports at either end of the connector. If the flow ports at the ends are atomic flow ports, then the types of those ports are often identical to the type of the item flow on the connector. If the flow ports at the ends are instead nonatomic flow ports, then they will be typed by a flow specification. The flow specification must contain a flow property whose type and direction match the item flow on the connector.

The IBD in Figure 4.7 is substantially similar to the one in Figure 4.4. This one, however, conveys an additional piece of information: an item flow that represents a ° C value flowing along the connector between the two nonatomic flow ports from the *eps* reference property to the *primaryComputer* part property. This item flow is compatible with these nonatomic flow ports, because the *Housekeeping Data* flow specification does, in fact, have a flow property of type ° C with a matching direction. (If you're skeptical, take a look at Figure 3.12 in Chapter 3.)

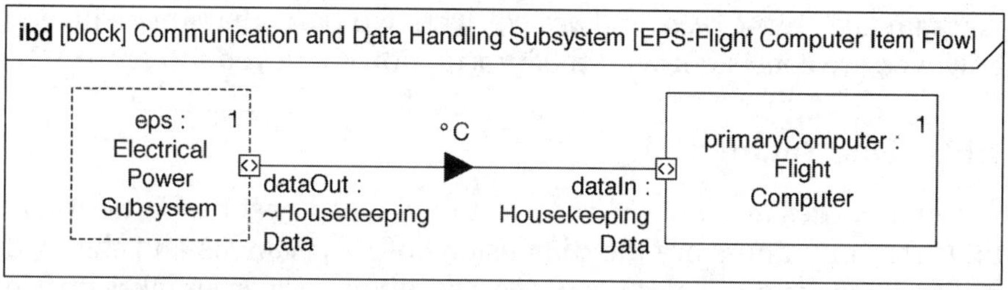

Figure 4.7 *An item flow on a connector*

4.10 Nested Parts and References

IBDs offer a powerful capability: displaying properties that are nested within other properties. Nesting enables you to convey multiple levels of the system hierarchy in a single view. This is necessary when your target audience needs to see the connections among nested parts. (I recommend you exercise this power judiciously, though; an IBD can quickly become unreadable.)

Figure 4.8 provides a sample IBD with nested properties. In this diagram, I've chosen to focus on the services that the part properties provide to one other. To show that, the IBD displays their standard ports and the interfaces assigned to those ports. This diagram shifts the focus to a different aspect of the system design from the one expressed in Figure 4.2, which displays flow ports to convey the types of things that flow among the properties.

Note

You can display standard ports as well as flow ports on the same IBD. My preference, though, is to focus on these two aspects of the system in separate diagrams.

The IBD in Figure 4.8 shows that the *DellSat-77 Satellite* block owns a part property named *cdhs* of type *Communication and Data Handling Subsystem*. The *cdhs* part property, in turn, owns part properties named *primaryComputer* and *backupComputer*.

This view of the model is consistent with the view shown in Figure 4.2. That IBD represents the *Communication and Data Handling Subsystem* block and displays all of its properties. In contrast, the IBD in Figure 4.8 omits several of the parts of *cdhs* that are not the focus of this diagram (*ant*, *tx*, *rx*, *mod*, and *demod*). The two parts that are displayed, however, are consistent with the information conveyed in Figure 4.2.

4.10.1 Dot Notation

SysML imposes no limit on how deeply you can nest properties on an IBD. The only limits are the dimensions of your canvas and the readability of the diagram. Nesting properties within properties takes up a lot of real estate on the diagram. SysML offers an alternative notation for conveying nested properties that overcomes this space constraint prob-

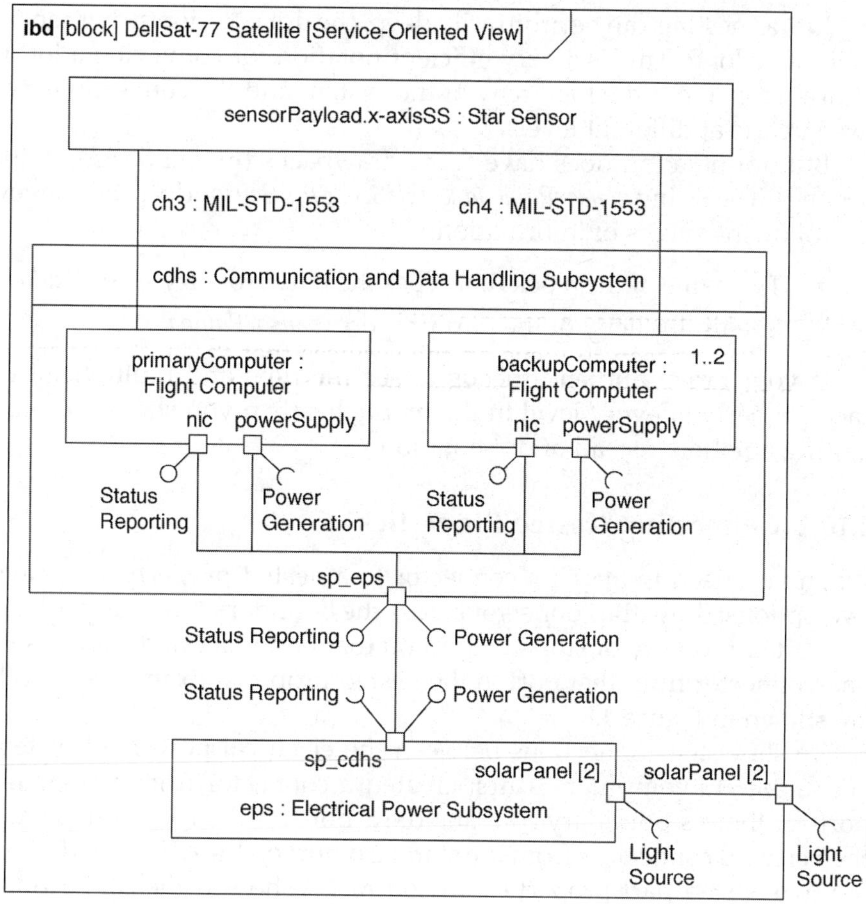

Figure 4.8 *An IBD with nested properties*

lem: dot notation. **Dot notation** enables you to express a structural hierarchy compactly in the form of a text string. An example is shown in the property at the top of the IBD in Figure 4.8. The string *sensorPayload.x-axisSS : Star Sensor* conveys several pieces of information:

- The *DellSat-77 Satellite* block owns a part property named *sensorPayload*.
- The part property *sensorPayload*, in turn, owns a property named *x-axisSS*.
- The property *x-axisSS* is typed by the block named *Star Sensor*.
- The multiplicity of *x-axisSS* is 1..1 (the default, because no multiplicity is shown).

Just as nesting can be arbitrarily deep, the dot notation string can be arbitrarily long. This is a very efficient notation for conveying a lot of information about the hierarchy of the system and the connections between parts at different levels.

But dot notation does have some drawbacks (in comparison with nesting). The string *sensorPayload.x-axisSS : Star Sensor* does *not* convey the following pieces of information:

- The name of the block that types the part property *sensorPayload*
- The multiplicity of the part property *sensorPayload*

If your target audience needs to see the type and multiplicity of each property at every level in the hierarchy, then you should use the nesting notation instead of dot notation.

4.10.2 Connecting Nested Properties

When you need to attach a connector to a nested property, you have two options: draw the connector across the boundary that encapsulates the nested property or stop at a port on that boundary and draw a second connector from that port to the nested property. Examples of both are shown in Figure 4.8.

To show the connections between the electrical power subsystem and the flight computers, I first created a connector from a standard port on the *eps* boundary to a standard port on the *cdhs* boundary. I then created connectors from the standard port on the *cdhs* boundary to the two nested part properties. In contrast, I showed the connections between the star sensor and the flight computers by creating connectors directly from the *x-axisSS* property to the nested *primaryComputer* and *backupComputer* properties, crossing the *cdhs* boundary in the process.

The decision to draw connectors across boundaries or stop at ports on the boundaries is a matter of judgment and should be based on knowledge of design principles. As mentioned in Chapter 3, ports let you specify blocks in a modular way, presenting an interface to a client that hides the internal implementation of a block. This is the object-oriented principle of **encapsulation**. And it's a good principle to adopt as your default mode of design.

When you draw a connector across a boundary, you're violating the principle of encapsulation. There are good reasons for doing this (such as satisfying performance constraints in a hard real-time embedded system), but those cases should be the exception rather than the rule. When you violate the principle of encapsulation, do it knowingly, and document your rationale in the model.

Summary

An IBD conveys an important aspect of a system's structure: the specific parts that will exist in a built system and the connections among those parts. This aspect of a system's design strongly complements the information you can convey on a BDD, and you often create these two kinds of diagrams in tandem.

An IBD has the unique ability to convey the services that specific parts provide to one another and the types of matter, energy, and data that can flow among them across their connections. This aspect of a system's structure is universally valued among system stakeholders.

Chapter 5

Use Case Diagrams

Specifying system use cases is a common design activity for systems engineering teams. The SysML use case diagram supports this activity. Use case diagrams let you display various kinds of elements and relationships to express information about the services your system provides and the stakeholders who require those services.

5.1 Purpose

A use case diagram concisely conveys a set of **use cases**—the externally visible services that a system provides—as well as the actors that invoke and participate in those use cases. A use case diagram is a black-box view of the system; it is therefore well suited to serve as a system context diagram.

5.2 When Should You Create a Use Case Diagram?

A use case diagram is an analysis tool and is generally created early in the system life cycle. System analysts may enumerate use cases and create use case diagrams during the development of the system concept of operations (ConOps). In some methodologies, analysts create use cases in lieu of text-based functional requirements during the requirements elicitation and specification stage of the system life cycle. System architects later analyze system-level use cases to derive and allocate subsystem- and component-level use cases during the architectural design stage.

5.3 Wait! What's a Use Case?

Before I discuss use case *diagrams* in detail, it's important for you to have a clear understanding of *use cases*. To jump-start this discussion, I defer to several highly authoritative sources.

In *The Unified Modeling Language Reference Manual*, second edition, James Rumbaugh, Ivar Jacobson, and Grady Booch define a use case as a "specification of sequences of actions, including variant sequences and error sequences, that a system, subsystem, or class can perform by interacting with outside objects to provide a service of value."

In *Writing Effective Use Cases*, p. 1, Alistair Cockburn explains it this way:

> A use case captures a contract between the stakeholders of a system about its behavior. The use case describes the system's behavior under various conditions as it responds to a request from one of the stakeholders, called the *primary actor*. The primary actor initiates an interaction with the system to accomplish some goal. The system responds, protecting the interests of all the stakeholders. Different sequences of behavior, or scenarios, can unfold, depending on the particular requests made and the conditions surrounding the requests. The use case gathers those different scenarios together.

Here are the key ideas to keep in mind when you're enumerating your system's use cases:

- A use case is a service—a behavior—that your system will perform. The use case name, therefore, is always a verb phrase (such as *Send command*).
- Not every behavior your system performs is a use case. Rather, use cases are the subset of system behaviors that external actors can directly invoke or participate in.
- An actor can be a person or an external system that interfaces with your system.
- The actors that invoke a use case are called **primary actors**. The actors that participate in the use case are called **secondary actors**. A primary actor can also be a secondary actor.
- Each use case should represent a primary actor's goal. Write the use case name—the verb phrase—from the perspective of the

actor and not your system. For example, if a satellite flight controller needs to send a command, then name the use case *Send command* and not *Receive command*.

- The use case name does not convey a lot of information. You will create a **use case specification** for each use case to narrate how your system and its actors collaborate to achieve the use case goal.

5.3.1 Use Case Specifications

The **use case specification** conveys the narrative that unfolds when a primary actor invokes the use case. Traditionally, use case specifications have been text documents. Alistair Cockburn offers an excellent use case specification format in his book *Writing Effective Use Cases*:

- **Use case name:** a verb phrase
- **Scope:** the entity that owns (provides) the use case (for example, the name of an organization, system, subsystem, or component)
- **Primary actor:** the actor that invokes the use case (the actor whose goal the use case represents)
- **Supporting (secondary) actors:** actors that provide a service to the system (participate in the use case by performing actions)
- **Stakeholder:** someone or something with a vested interest in the behavior of the system
- **Preconditions:** the conditions that must be true for this use case to begin
- **Guarantees (postconditions):** the conditions that must be true at the end of the use case
- **Trigger:** the event that gets the use case started
- **Main success scenario:** the scenario (the sequence of steps) in which nothing goes wrong
- **Extensions (alternative branches):** alternative sequences of steps branching off of the main success scenario
- **Related information:** whatever your project needs for additional information

You can also create graphical use case specifications using SysML activity diagrams. **Activity diagrams** tend to be more concise and less ambiguous than the traditional text form of a use case specification. With that said, many modelers make the mistake of jumping right into a modeling tool to create the activity diagram while brainstorming the use case narrative. This practice tends to stifle creativity. You need to focus first on the narrative; write a good story that clearly and correctly conveys how actors will interface with your system when it executes the use case. Once you begin creating an activity diagram, you will spend much of your time and energy on the *layout* of the activity diagram, and that will detract from your ability to tell a clear and correct story. I always advise the designers on my team to capture the narrative in text form first (using the template shown above), refine the narrative with the stakeholders, and *then* jump to the modeling tool to express the narrative visually on an activity diagram. (Read more on activity diagrams in Chapter 6, "Activity Diagrams.")

Be forewarned: It's typical, on the first try, to create use cases that are too broad and too high level. If you do that, you'll discover the problem when you try to write the use case specification. The narrative will have many decision points and many paths of execution that you regard as nominal (success) scenarios. Consider this a red flag. You need to refine each use case either by using more precise verbs or by including qualifiers in the predicate to create a set of more finely grained use cases. For example, the narrative for *Send command* will likely be too broad, because it encompasses several different ways to generate the command and multiple channels for transmission, all of which are simply variations of success scenarios. You could refine that use case to create the following set of use cases:

- *Send real-time command via uplink*
- *Send real-time command via forward link*
- *Send stored command via uplink*
- *Send stored command via forward link*

You will know that a use case is at an appropriate level of granularity when it has exactly *one* main success scenario. All other paths of execution through the use case specification should be error or exception sequences.

5.3.2 Use Cases versus Scenarios

Use case and *scenario* are not synonyms. Each path of execution through a use case from beginning to end is a distinct **scenario**. A use case therefore consists of one or more scenarios. At a minimum, a use case consists of a main success scenario—the nominal path of execution. Often it also contains additional scenarios representing error and exception sequences, which branch off of the main success scenario.

A SysML sequence diagram is well suited to graphically represent a single scenario. A common technique is to dissect a single use case specification—either a text specification or an activity diagram—to create a *set* of sequence diagrams, one per scenario. Those sequence diagrams can then pull double duty as test cases if you take the additional step of specifying values for the inputs and the expected outputs (more on sequence diagrams in Chapter 7, "Sequence Diagrams").

5.4 The Use Case Diagram Frame

The diagram kind abbreviation for a use case diagram is *uc*. The model element type that the diagram frame represents can be any of the following:

- *package*
- *model*
- *modelLibrary*
- *view*

The name of the use case diagram in Figure 5.1 is "System Use Cases." The diagram header also tells us that this diagram represents the *Behavior* package in the system model. The *Behavior* package, therefore, is the namespace for the use cases shown on the diagram. That is equivalent to saying that the *Behavior* package contains the use cases shown on the diagram.

Of course, I could have organized this model differently. I could have chosen to segregate the system's use cases into separate packages with names that correspond to the stages of the mission life cycle: a *Manufacturing Use Cases* package, a *Launch Use Cases* package, and an *Operations Use Cases* package. In that case, I would have created a separate use case diagram for each of the packages and changed the model element name in the header for each one accordingly. I also would have given each diagram a unique name to convey which aspect of the model was in focus.

I say all this simply to point out that there are many ways to organize your model. The key idea is that the organization of your model generally affects what appears in the header and the contents area of each diagram you create.

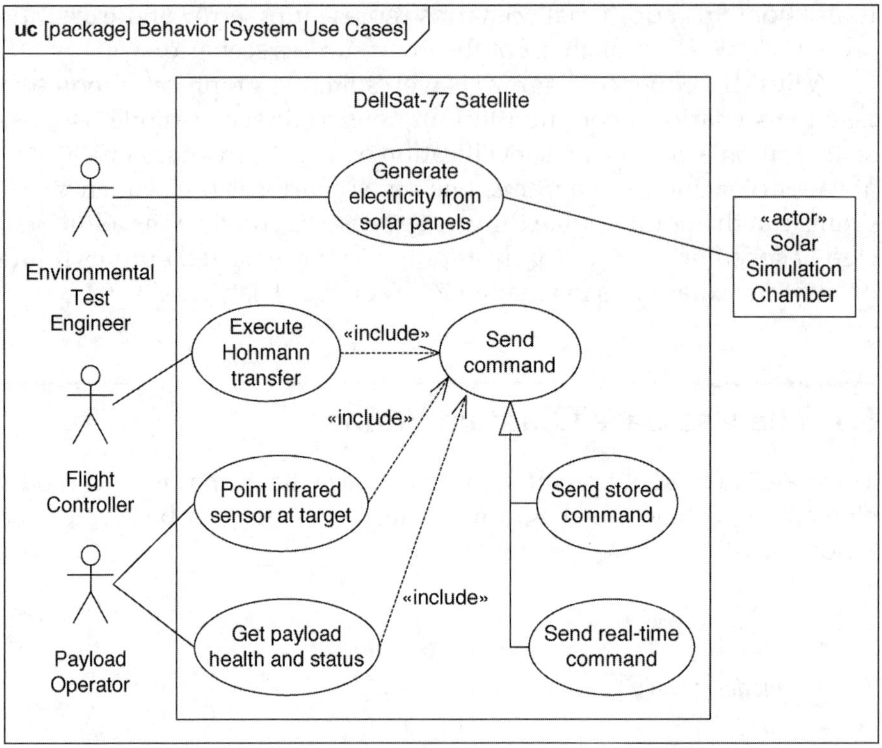

Figure 5.1 *A sample use case diagram*

5.5 Use Cases

The notation for a use case is an ellipse (oval). You can display the name of a use case—generally, a verb phrase—either inside or beneath the ellipse. (Putting the name inside the ellipse is much more common.)

As with a block, a use case can be generalized and specialized, and this means that you can create and display generalization relationships from one use case to another. A generalization means here exactly what it means in the context of blocks: inheritance. As mentioned in Chapter 3, "Block Definition Diagrams," you read this relationship in English as "is a type of." The notation for a generalization is the same as in Chapter 3: a solid line with a hollow, triangular arrowhead on the end of the generalized element, the supertype. The specialized element, the subtype, appears at the tail end of the line.

Figure 5.1 displays a generalization from the *Send stored command* use case to the *Send command* use case. The *Send stored command* use case is a subtype of the *Send command* use case. The same is true of the *Send real-time command* use case. These two subtypes inherit everything the supertype has, including its relationships with other use cases and with actors as well as any behaviors (activities, interactions, or state machines) that may be associated with it. The subtypes may optionally redefine what they inherit with more specific definitions or add new relationships and behaviors that the supertype doesn't have.

5.6 System Boundary

The **system boundary** (also called the **subject**) represents the system that owns and performs the use cases on the diagram. The notation for the system boundary is a rectangle that encloses the use cases (not to be confused with the diagram frame). The name of the subject—shown at the top inside the rectangle—must always be a noun phrase. In Figure 5.1, shown earlier, the system boundary is DellSat-77 Satellite.

During ConOps development, the name of the system boundary will be the name of the system you're designing. At that stage, your goal is to represent the system as a black box and specify the services it will provide to the actors in its environment.

Later, during architectural design, you will structurally decompose the system into subsystems (possibly) and components (definitely). At that stage, you will create new use case diagrams for each subsystem and eventually for each component. When you do, the name of the system boundary will be the name of the subsystem or component you're working on. (And if it bothers you to refer to it as the "system" boundary at that point, feel free to call it the "subject" instead.)

5.7 Actors

As discussed in Section 3.8, "Actors," there are two notations for an actor: (1) a stick figure or (2) a rectangle with the keyword «actor» preceding the name. I use both notations in Figure 5.1. Recall that either notation is legal for any type of actor—a person or a system. However, modelers typically adopt the convention of using the stick figure notation to represent a person and the rectangle notation to represent a system.

As with a BDD, you can display generalizations among actors on a use case diagram. As before, it means that a subtype (at the tail end) inherits all the structural and behavioral features of its supertype (at the triangular arrowhead end). If the supertype has an association with a use case, then the subtype also inherits that association and has access to that use case.

5.8 Associating Actors with Use Cases

You would show that an actor interacts with your system to invoke or participate in a use case by creating an association between the actor and the use case. I discuss associations in detail in Chapter 3. They mean the same thing on use case diagrams: A link could exist between instances of the associated elements during system operation; an instance of an actor may invoke or participate in an instance of a use case (i.e., an **execution** of a use case).

Recall that multiplicity represents a constraint on how many instances can be involved in a relationship at any given time. Figure 5.2 shows a multiplicity of 1..* on the *Ground Station Radar* end of an association and a multiplicity of 1..2 on the *Radar Operator* end of an association. This conveys that a single execution of the *Track satellite trajectory* use case can involve one or more ground station radar units and either one or two radar operators.

This figure also shows a multiplicity of 0..* on the use case end of the two associations. This conveys that any particular radar operator or ground station radar unit may be involved in zero or more executions of *Track satellite trajectory* at any given moment.

Be careful to avoid the following illegal things with associations on use case diagrams:

- You cannot create a composite association between an actor and a use case.
- You cannot create an association (of any kind) between two actors.
- You cannot create an association (of any kind) between two use cases.

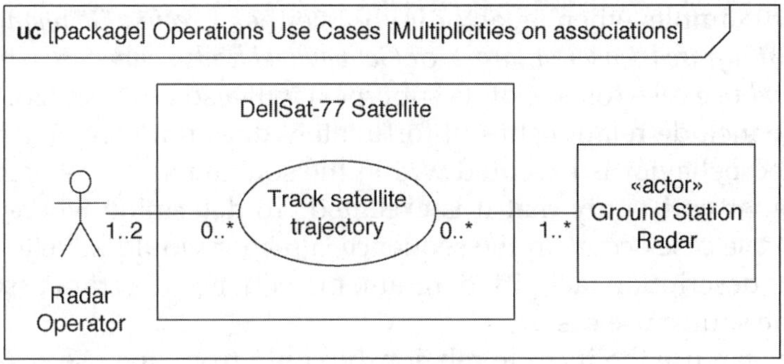

Figure 5.2 *Associations with multiplicity specified*

5.9 Base Use Cases

A **base use case** is any use case that is connected to a primary actor via an association relationship. This means that a base use case is one that represents a primary actor's goal.

Four base use cases are shown in Figure 5.1: *Generate electricity from solar panels, Execute Hohmann transfer, Point infrared sensor at target,* and *Get payload health and status*. The other three use cases are included use cases, as discussed in the next section.

5.10 Included Use Cases

An **included use case** is any use case that is the **target**—the element at the arrowhead end—of an **include relationship**. The notation for an include relationship is a dashed line with an open arrowhead and the keyword «include» floating next to it. In Figure 5.1, the *Send command* use case is an example of an included use case.

Despite the similarity in the notations for an include relationship and a dependency relationship, an include relationship is *not* a kind of dependency; the client–supplier semantics of a dependency do not apply here. The include relationship instead conveys that when the use case at the *source* end—the tail end of the relationship—gets invoked, the included use case at the target end is also executed. Stated differently, the included use case behavior is a required part of the use case at the source end.

For example, when an actor invokes either *Execute Hohmann transfer, Point infrared sensor at target*, or *Get payload health and status*, the *Send command* use case (or one of its subtypes) will also be executed.

The include relationship unfortunately does not convey *where* the included behavior is executed within the source use case—beginning, middle, or end—only that it is executed. To determine where the included use case occurs in the sequence, a reader would have to look at the text description, activity diagram, or sequence diagram associated with the source use case.

You can use the include relationship only from one use case to another. It is generally drawn from a base use case—a use case associated with a primary actor—to an included use case.

Note

It is legal to draw an include relationship from one included use case to another (thus creating a chain of included use cases), but it's best to avoid this practice. I've seen beginning modelers create unreadable use case diagrams in the process.

You should create an included use case only when you want to represent a common chunk of behavior that you've factored out of several base use cases. Many beginning modelers make the mistake of using include relationships gratuitously to show a functional decomposition of a single base use case. That is not the intended purpose of the include relationship.

As a rule, I don't create included use cases in my first iteration on a use case model. I begin by brainstorming the base use cases, which represent the primary actors' goals. Then, with input from the system stakeholders, I begin writing the text narratives for those use cases. A pattern then begins to emerge. I will notice that several base use cases perform a common step (or steps); that is when I factor out that chunk of common behavior and represent it as an included use case.

One last pitfall to avoid: Never create an association relationship between a primary actor and an included use case. A primary actor should be associated only with the base use cases that represent its goals. For example, the flight controller's goal is to execute a Hohmann transfer and not to send a command. Sending a command is simply a necessary step along the way to accomplishing the actor's goal.

5.11 Extending Use Cases

An **extending use case** is any use case that is the source—the element at the tail end—of an **extend relationship**. The notation for an extend relationship is a dashed line with an open arrowhead and the keyword «extend» floating next to it. In Figure 5.3, the *Switch to TDRS telemetry feed* use case is an example of an extending use case.

Like an include relationship, an extend relationship is *not* a kind of dependency, despite the similarity in their notations. Rather, the extend relationship conveys that when the use case at the target end—the arrowhead end of the relationship—gets invoked, the extending use case at the source end may *optionally* be executed, too. This means that the use case at the target end of the extend relationship is complete by itself.

For example, each time a radar operator invokes the *Track satellite trajectory* use case, the *Switch to TDRS telemetry feed* use case may also be executed—or it may not; execution of that extending use case depends on whether some triggering condition in the *Track satellite trajectory* use case is satisfied.

You can specify that triggering condition in a comment anchored to the extend relationship. In most cases, though, I prefer to specify the

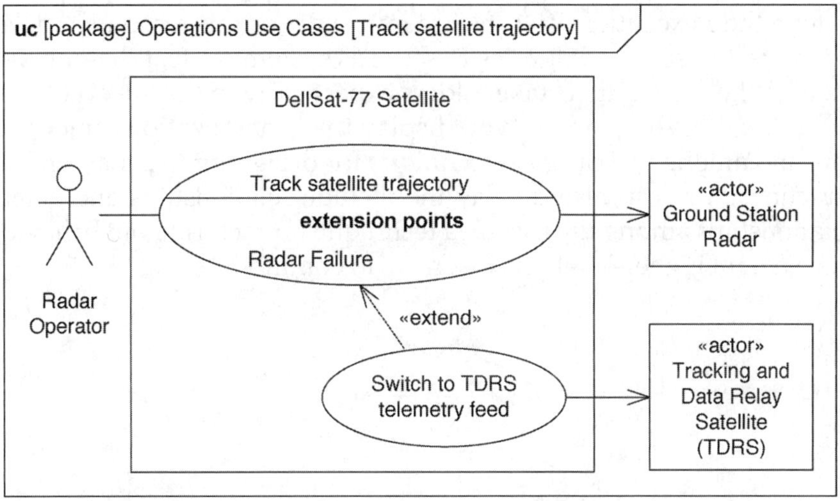

Figure 5.3 *An extending use case*

triggering condition in the activity diagram that I create for the target use case rather than clutter the use case diagram itself.

You can also specify the extension point where the extending use case branches off within the behavior of the target use case. That extension point is named in a compartment of the target use case.

My advice, though, is to do this sparingly. I generally like to keep my use case diagrams as clean as possible and specify all the details of the extending use case (the extension point and the triggering condition) on the activity diagram associated with the target use case. The activity diagram will contain a decision node that corresponds to the extension point. The extending use case behavior branches off from that decision node.

In the preceding section, I state that I never create included use cases in my first iteration on a use case model. The same is true about extending use cases. As mentioned, I begin by writing the text specifications for the base use cases. When I notice that several base use cases contain the same optional behavior, I factor out that chunk of common behavior and represent it as an extending use case.

Summary

A use case diagram presents a black-box view of the services that a system provides and the actors who require those services and participate in their execution after they're invoked. For this reason, modelers often create use case diagrams to serve as system context diagrams—a view of the system that stakeholders expect early in the life cycle.

A use case diagram lets you display the generalizations among actors and among use cases—a technique for designing to behavioral abstractions. You can also display the include relationships and extend relationships among use cases—a technique for factoring out behaviors that several higher-level services have in common.

Chapter 6

Activity Diagrams

Activity diagrams are one of three kinds of SysML diagrams that you can use to express information about a system's dynamic behavior. An activity diagram can display various kinds of actions, enabling you to convey even the most complex behavioral narratives. Object nodes let you model the flow of matter, energy, and data through an activity, and you can use control nodes to steer the execution of an activity. Activity partitions enable you to allocate system behaviors to system structures.

6.1 Purpose

An **activity diagram** is a kind of behavior diagram; it's a dynamic view of the system that expresses sequences of behaviors and event occurrences over time. This is in contrast to structure diagrams (BDDs, IBDs, and parametric diagrams), which are static views that convey no sense of the passage of time or change within the system and its environment.

Activity diagrams, sequence diagrams, and state machine diagrams are the three options that SysML offers you to specify system behavior. All three can express sequential and concurrent behaviors and event occurrences over time. However, each one has strengths and weaknesses that make it more or less appropriate based on the needs of your target audience.

An activity diagram is particularly good at expressing the flow of objects—matter, energy, or data—through a behavior, with a focus on how the objects can be accessed and modified in an execution of that behavior during system operation. One of its key advantages is its readability; activity diagrams can express complex control logic better than sequence diagrams and state machine diagrams. And activity diagrams are uniquely capable of expressing continuous system behaviors.

Activity diagrams do have a disadvantage: moderate ambiguity. Activity diagrams express the *order* in which actions are performed, and they can optionally express which structure *performs* each action. They do not, however, offer any mechanism to express which structure *invokes* each action. (Sequence diagrams, in contrast, can express all three pieces of information.)

For these reasons, modelers typically use activity diagrams as analysis tools when working with stakeholders to define the problem space and specify the required behavior of the system. Activity diagrams are not particularly useful tools for detailed design—that is, unambiguous specifications of behavior suitable for system implementation.

6.2 When Should You Create an Activity Diagram?

Because an activity diagram is most effective as an analysis tool, it often is the first kind of behavior diagram you turn to when you need to communicate with stakeholders and capture the expected behaviors of the system and its actors. You also use activity diagrams to communicate with other members of your team and capture the expected behaviors of the system's internal parts. In short, creation of activity diagrams is not tied to any particular stage of the system life cycle.

Some modeling methodologies prescribe that you create activity diagrams to serve as graphical use case specifications (either instead of or in addition to the text specifications discussed in Chapter 5, "Use Case Diagrams"). If you adopt this practice, you will create activity diagrams whenever you add new use cases to your system model.

6.3 The Activity Diagram Frame

The diagram kind abbreviation for an activity diagram is *act*. The only allowable model element type for an activity diagram is *activity*. The

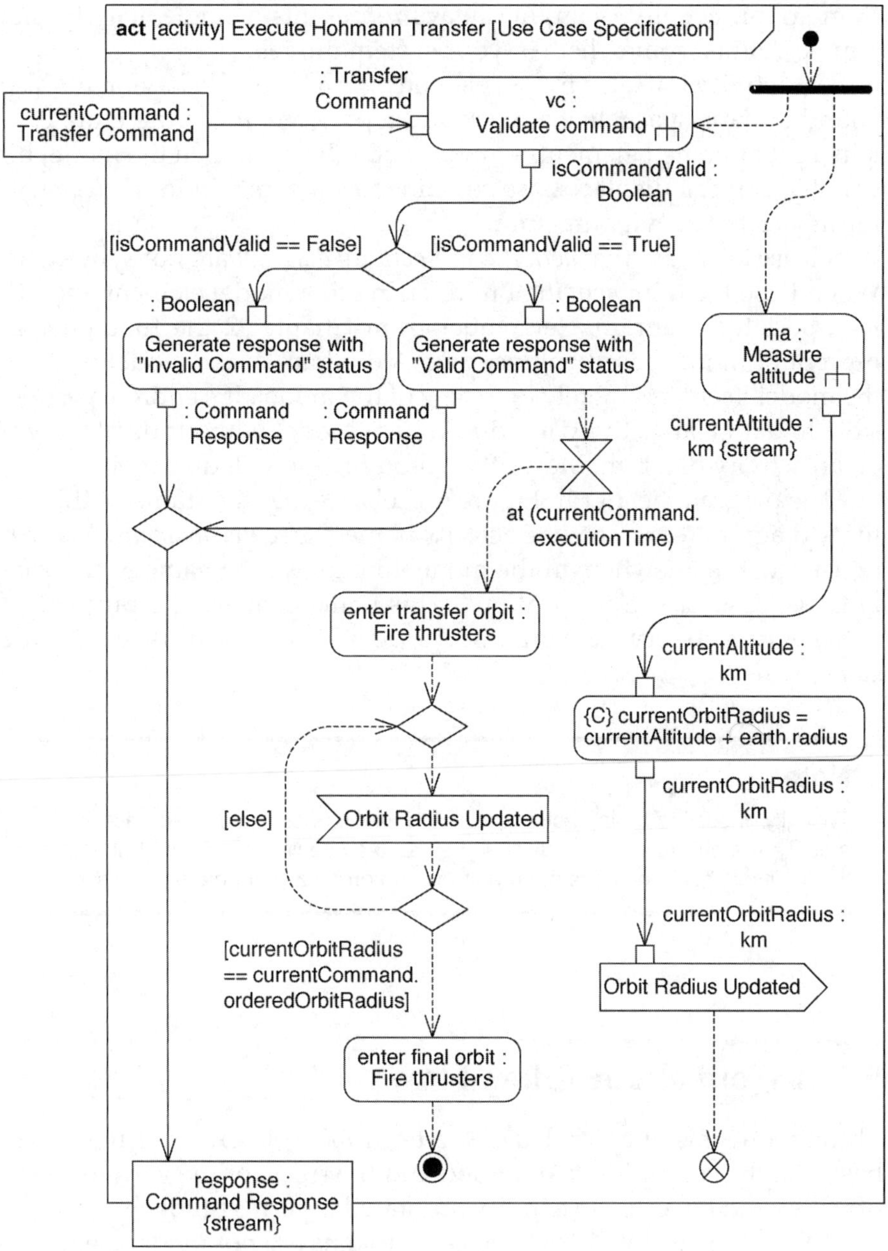

Figure 6.1 *A sample activity diagram*

frame of an activity diagram always represents a single activity that you've defined somewhere in your system model.

An **activity** is itself a model element; it's a kind of behavior. It's also a kind of namespace, like a block and a package. It can therefore contain a set of named elements—nodes and edges—within the model hierarchy. You can display those contained elements within the frame of the associated activity diagram.

Please be aware that *activity* and *activity diagram* are not synonyms. When I use the term *activity*, I'm referring to a model element and not its associated diagram. Remember from Chapter 2 the fundamental precept of model-based engineering: A diagram of the model is never the model itself; it is merely one view of the model. It's entirely permissible to define an activity in your system model without displaying it on an activity diagram. Most often, though, you will display it.

The diagram header in Figure 6.1 tells us that the frame of this activity diagram represents the activity named *Execute Hohmann Transfer*, which exists somewhere in the model hierarchy. The name of this diagram is "Use Case Specification." This name conveys the purpose of this diagram: to serve as a graphical specification for the use case of the same name.

Note

Typically, modeling tools nest an activity under its associated use case in the model hierarchy when the activity serves as the specification for that use case. SysML itself, however, does not require you to organize your model in this way.

6.4 A Word about Token Flow

Throughout this chapter, I use the term *token*, because activities are based on the concept of token flow. And the rules for the various kinds of nodes and edges in an activity are stated in terms of token flow.

Token flow is an abstract concept. Tokens are not model elements. You don't create them in your system model, and you don't display them on activity diagrams. The idea of tokens is inherent in the meaning that SysML defines for activities, nodes, and edges. You must rely on your imagination to envision the flow of tokens through an activity based on the rules for each kind of node and edge.

I find it helpful to think of a token as a Monopoly game piece moving across the activity diagram, traversing edges from one node to the next. There can be multiple tokens flowing through a single execution of an activity. Each token moves independently of the others based on the type and state of the token itself, the actions and control logic that define the activity, and any events that occur during an execution of the activity. (All this will become clearer in the sections that follow. I promise.)

What does a token represent? The answer comes in two parts. First, there are two kinds of tokens: object tokens and control tokens. I'll start with the easier one.

An **object token** represents an instance of matter, energy, or data that flows through an activity. It can represent an input or an output of the activity as a whole, and it can represent an input or an output of an action within the activity. Formally, an object token represents an instance of a block, value type, or signal that you've created somewhere in your model hierarchy (to define a type of matter, energy, or data). As you might expect, there can be multiple object tokens flowing through a single execution of an activity.

A **control token** represents . . . nothing. Nothing physical, that is. It has no type (block, value type, or signal). A control token simply indicates which action in an activity is currently enabled at a particular moment during an execution of the activity. Potentially, there could be multiple control tokens flowing through a single execution of an activity. If so, it simply means that multiple actions in the activity are enabled concurrently.

I know that this is rather abstract stuff at this point, but with this foundation in place at the beginning, you'll be able to make better sense of the explanations and examples that follow in this chapter. As I discuss each kind of element that can appear on an activity diagram, I explain the rules for that element in terms of token flow. And there are a lot of rules; it may seem overwhelming at first. But as you practice reading and creating activity diagrams, you will synthesize the rules into a meaningful whole that will enable you to interpret and express even the most complex system behaviors.

6.5 Actions: The Basics

An **action** is one kind of node that can exist within an activity; it's a node that models a basic unit of functionality within the activity. An

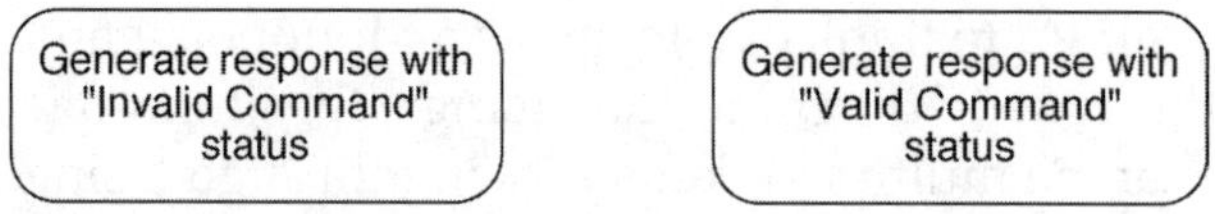

Figure 6.2 *Actions with natural language expressions*

action represents some form of processing or transformation that will occur when the activity gets executed during system operation.

The notation for a basic action is a round-cornered rectangle (colloquially called a **round-angle**). I say "basic" because there are several specialized kinds of actions, each with its own notation (more on that in Section 6.8).

You can enter any behavioral description you like in an action; your description gets displayed as a string inside the round-angle on an activity diagram. Most often, system modelers write actions as verb phrases expressed in a natural language (such as English). Figure 6.2 shows examples.

Although SysML does not require it, I recommend as a best practice that you always write an action as a phrase that begins with a strong, unambiguous verb. Also, you should avoid putting multiple verbs into a single action; break it up into several atomic actions instead. For example, instead of a single action, "Validate and save command," create two sequential, atomic actions: "Validate command" and "Save command."

As an alternative to natural languages, you can use formal languages (such as C, Java, Verilog, or Modelica) for an action in an activity. SysML calls such a statement an **opaque expression**. This jargon isn't important, but you should know that an opaque expression has two parts: a language and a body. You specify the **language** in curly brackets preceding the **body**. Figure 6.3 shows an example of an action with an opaque expression written in the C programming language. (But note that opaque expressions are not limited to programming language statements.)

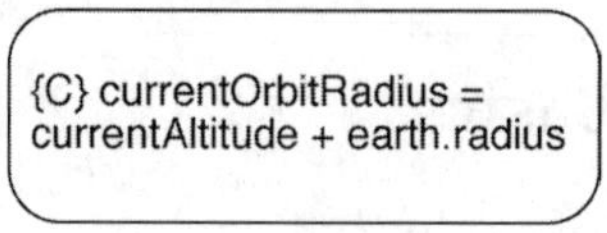

Figure 6.3 *An action with an opaque expression*

System modelers seldom write actions as opaque expressions. Development teams do this more often in the activity diagrams they create to communicate their designs. But, as a stakeholder to those development teams, you should be prepared to interpret activity diagrams that contain opaque expressions when you see them.

A useful and meaningful activity always consists of more than one action (as shown in Figure 6.1). You connect the actions in an activity by using edges that define ordered (and sometimes concurrent) sequences. Those sequences convey the larger narrative of the activity as a whole. Like the text narratives described in Chapter 5, a good activity diagram tells a clear story.

The story you tell using an activity diagram can convey more than sequences of actions; it can also convey the flow of objects—the inputs and outputs of those actions and of the activity as a whole. (You can read more on this in the next section.)

6.6 Object Nodes

An **object node**, another kind of node that can exist within an activity, models the flow of object tokens through an activity (where each object token, again, represents an instance of matter, energy, or data). An object node most often appears between two actions to convey that the first action produces object tokens as outputs, and the second action consumes those object tokens as inputs.

The notation for an object node is a rectangle. The name string that appears inside an object node has the following format:

```
<object node name> : <type> [<multiplicity>]
```

The object node name is modeler defined. The type must match the name of a block, value type, or signal that you've defined somewhere in your model hierarchy; it specifies the nature of the object tokens that the object node can hold. The multiplicity specifies how many object tokens the object node can hold at any given moment during an execution of the activity. If not shown in the name string, the default multiplicity for an object node is 1..1.

The activity fragment in Figure 6.4 displays an example of an object node named *currentAltitude*. It holds object tokens that represent instances of the value type *km*. The multiplicity shown at the end of the string conveys that the first action will produce exactly one object token

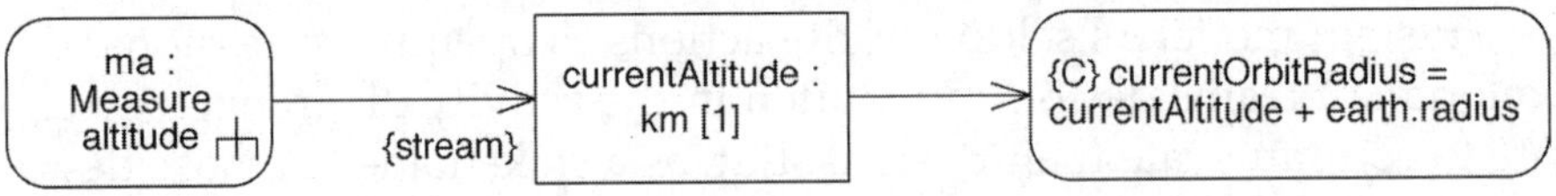

Figure 6.4 *An object node between two actions*

as an output, and the second action requires exactly one object token as an input (in order for it to start).

Another useful feature of an object node is that it can optionally display compartments—just as a block or a part property can—to convey the internal properties of the object tokens it holds. Of course, this is useful only when the specified type is a block, value type, or signal that actually owns internal properties. A primitive value type, such as *km*, does not.

6.6.1 Pins

A **pin** is a specialized kind of object node. You attach a pin to an action to represent an input or output of the action. The notation for a pin is a small square attached to the boundary on the outside of an action, as shown in Figure 6.5.

You can optionally display an arrow inside the square to specify whether the pin represents an input or an output. This is redundant, though, if you join two pins with an edge, and most of the time you will. The format of the name string is the same for a pin as for an object node, but it floats near the pin instead of being displayed inside the pin (for obvious reasons).

The activity fragment in Figure 6.5 is an alternative view of the very same model shown in Figure 6.4. Pins mean exactly what object nodes mean; pins are simply alternative notations that you can choose case-by-case to meet the specific needs of your target audience. Each has a strength and a weakness.

The object node notation lets you display compartments to convey the internal properties of the object tokens it holds. However, it takes

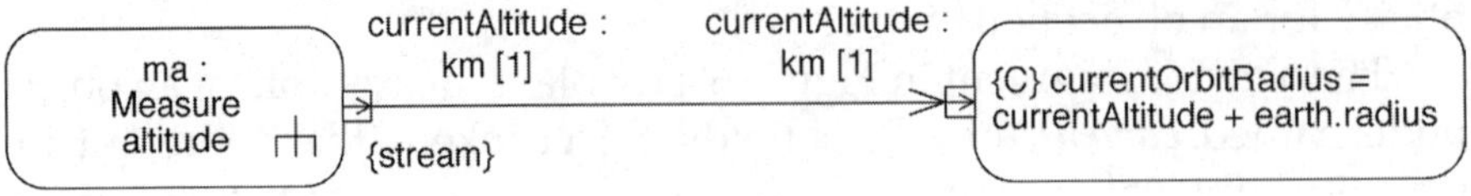

Figure 6.5 *Actions with pins*

Figure 6.6 *An action with optional pins*

up more real estate on an activity diagram than the pin notation. The pin notation does not allow you to display compartments, but it's a much more space-efficient notation. As a rule, I recommend that you adopt the pin notation as your default option and switch to the object node notation only on those rare occasions when you need to show the internal properties of an object token.

Figure 6.6 shows an example of an input pin and an output pin, each having a lower multiplicity of zero. This is how you would model an action that has optional inputs or outputs. An action with an optional input pin can start even with no object tokens at that pin. An action with an optional output pin can execute and possibly produce no object tokens at that pin. Although it's redundant, SysML requires you to apply the «optional» stereotype to a pin (preceding the name) when its lower multiplicity is zero.

6.6.2 Activity Parameters

An **activity parameter** is another specialized kind of object node. You attach it to the frame of an activity diagram to represent an input or an output of the activity as a whole. The notation for an activity parameter is a rectangle straddling the frame of an activity diagram, as shown in Figure 6.7. The format of the name string is the same for an activity parameter as for an object node (and a pin).

SysML does not dictate where you must place an activity parameter on the frame of the diagram. A common modeling convention, however, is to place *input* activity parameters either on the top or left side of the frame and *output* activity parameters either on the bottom or right side of the frame. In fact, the only definitive way to tell the difference is the direction of the edge that's attached to an activity parameter.

Like a pin, an activity parameter can have a lower multiplicity of zero. This is how you would model an optional parameter for the activity as a whole. An activity with an optional input activity parameter

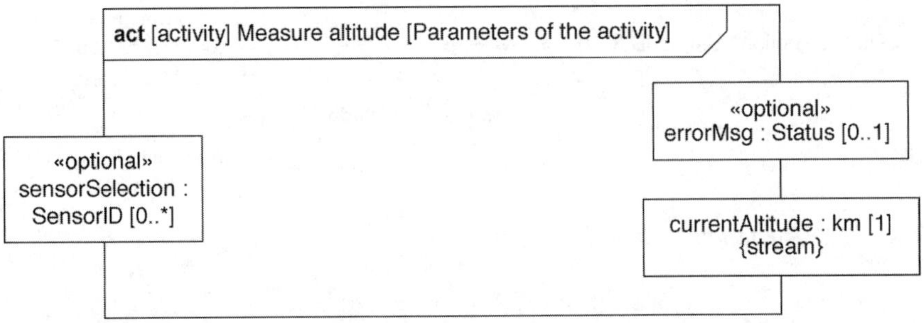

Figure 6.7 *An activity diagram frame with activity parameters attached*

(e.g., *sensorSelection*) can start even with no object tokens at that parameter. An activity with an optional output activity parameter (e.g., *errorMsg*) can execute and possibly return no object tokens via that parameter to the client that invoked the activity. As with pins, SysML requires you to apply the «optional» stereotype to an activity parameter (preceding the name) when its lower multiplicity is zero.

Given the examples in Figures 6.6 and 6.7, you've likely inferred that there's a relationship between pins and activity parameters. I discuss that relationship in more detail in Section 6.8, "Actions, Revisited."

6.6.3 Streaming versus Nonstreaming

By default, actions and activities consume their input object tokens only at the moment they begin executing. Similarly, they deliver their output object tokens only at the moment they finish executing. We refer to this as **nonstreaming** behavior.

However, the systems you design won't always behave this way; sometimes they will receive inputs and produce outputs even while behaviors continue to execute. We refer to this as **streaming** behavior. You can model streaming behavior by specifying *{stream}* at the end of the name string for a pin or an activity parameter.

Let's look at the nonstreaming case first. Figure 6.8 shows an action with a nonstreaming input pin and a nonstreaming output pin. When

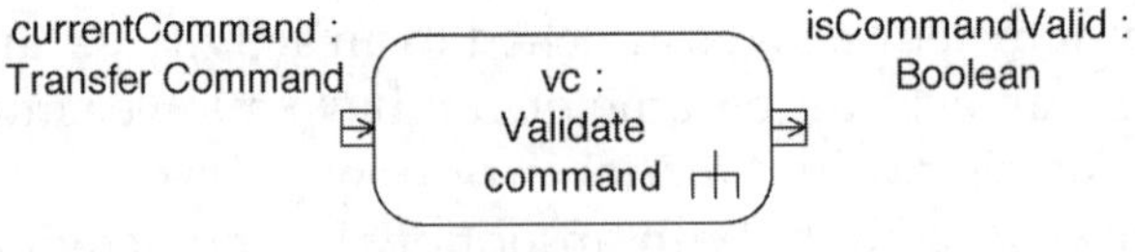

Figure 6.8 *An action with nonstreaming pins*

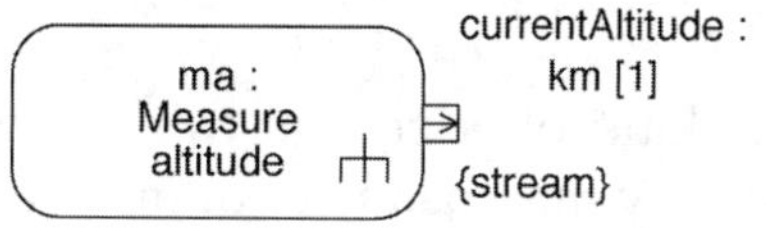

Figure 6.9 *An action with a streaming pin*

an object token of type *Transfer Command* arrives at the input pin *currentCommand*, the action *vc* will begin executing and consume that object token. If a second instance of *Transfer Command* arrives at the input pin while *vc* is currently executing, it will not get consumed until *vc* completes and then starts a second time.

When *vc* executes, it will internally generate an object token of type *Boolean*, but that object token will not get posted to the output pin *isCommandValid* until *vc* finishes executing. Thus, any actions that follow *vc* (and require that *Boolean* value as an input) cannot begin until *vc* completes.

That's the nonstreaming case. Now let's look at the streaming case. Figure 6.9 shows an action, *ma*, with a streaming output pin, *currentAltitude*. When *ma* executes, it will internally generate an object token of type *km*. That object token will get posted to the output pin even while *ma* is executing. Thus, any actions that follow *ma* (and require that *km* value as an input) can begin even while *ma* continues to execute (and potentially produce additional object tokens). This is one way to model two or more actions that execute concurrently.

Streaming has a similar meaning on the input side of an action. Object tokens that arrive at a streaming input pin are immediately available to the action even if it's already executing due to the arrival of object tokens earlier.

Streaming and nonstreaming mean the same thing in the context of activity parameters as they do for pins, except that the rules apply to the activity as a whole. Streaming pins and activity parameters enable you to model continuous system behaviors.

6.7 Edges

Activities can contain two general kinds of elements: nodes and edges. In the previous sections, I introduce two kinds of nodes: actions and object nodes. In this section, I introduce two kinds of **edges** that you can use to connect nodes to form ordered sequences in an activity: object flows and control flows.

6.7.1 Object Flows

An **object flow** is the kind of edge that transports object tokens. You use object flows to convey that instances of matter, energy, or data flow through an activity from one node to another when the activity executes during system operation.

The notation for an object flow is a solid line with an open arrowhead. Object flows generally connect two object nodes to each other.

Note

Pins and activity parameters are kinds of object nodes; they are implicitly included in this discussion.

In addition to an object node, though, you can have a decision node, merge node, fork node, or join node at one end of an object flow to di-

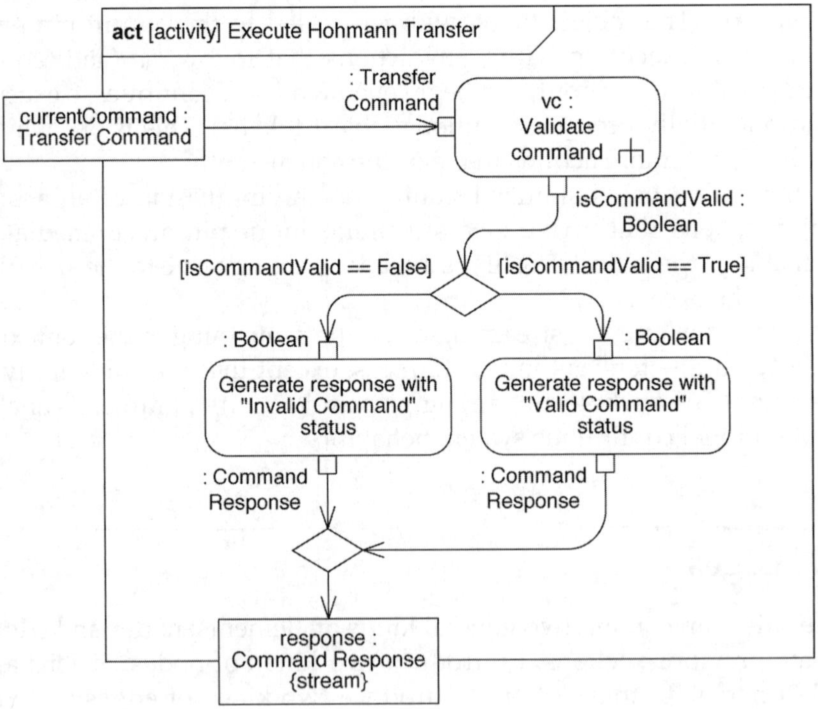

Figure 6.10 *Object flows*

rect the flow of the object tokens. I discuss these in detail in Section 6.9, "Control Nodes."

Figure 6.10 displays an excerpt of the larger activity diagram shown earlier in Figure 6.1. This excerpt shows seven examples of object flows. They convey that the actions in this activity fragment require objects as inputs and produce objects as outputs.

You must ensure that the object nodes at the ends of an object flow have compatible types; the object token produced as an output on the tail end must be acceptable as an input on the arrowhead end. You can satisfy this constraint in one of two ways:

- The types can be identical (as shown in Figure 6.10).
- The upstream type can be a subtype of the downstream type (as shown in Figure 6.11).

The activity diagram fragment in Figure 6.11 shows a valid variant of the behavior displayed in Figure 6.10. *Transfer Command* is a subtype of *Command* (as conveyed by the generalization relationship between them, which is displayed in the BDD). And generalization relationships imply substitutability (meaning that the subtype will be accepted wherever its supertype is required). The action *vc* requires an input of type *Command*. It will therefore accept any of its subtypes, including an instance of *Transfer Command*.

This is an example of designing to an abstraction (e.g., *Command*). You should become comfortable with this practice; it creates extensibility in your system design, and that minimizes the cost of changes that inevitably occur later in the life cycle.

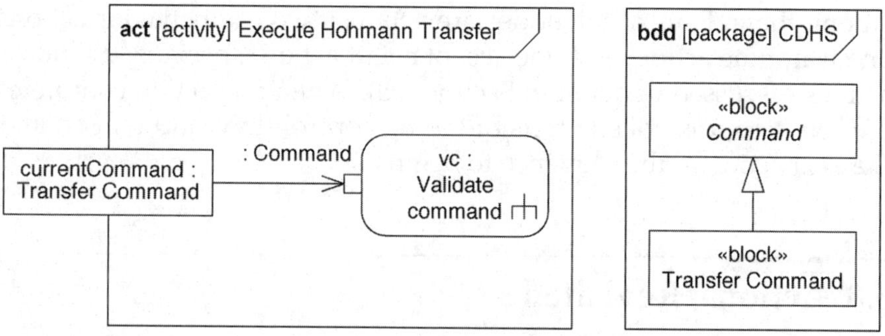

Figure 6.11 *Ensuring compatibility via generalizations*

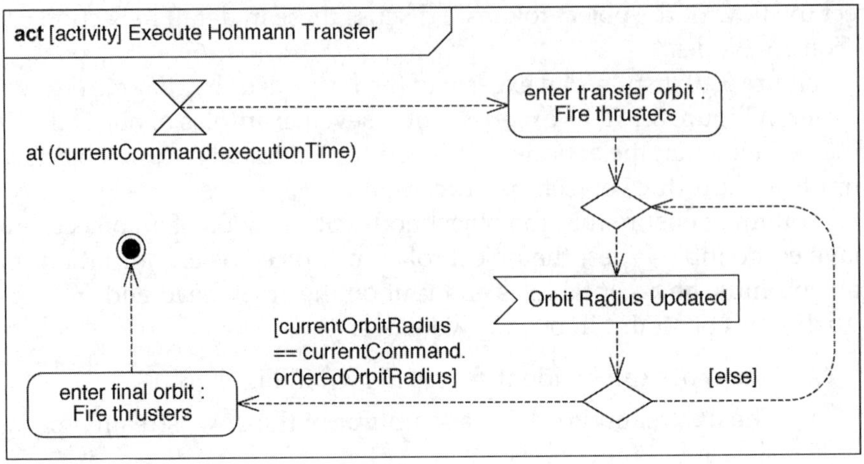

Figure 6.12 *Control flows*

6.7.2 Control Flows

A **control flow** is the kind of edge that transports control tokens. And the arrival of a control token enables an action that's waiting for one. You therefore use control flows to convey sequencing constraints among a set of actions when the object flows in your activity do not by themselves convey the intended sequence.

SysML allows two notations for a control flow: a dashed line with an open arrowhead or a solid line with an open arrowhead. I recommend that you stick with the dashed-line option if your modeling tool supports it. In that way, your readers can easily distinguish control flows from object flows in your activity diagrams.

Figure 6.12 displays an excerpt of the larger activity diagram shown in Figure 6.1. This excerpt shows seven examples of control flows, which connect the nodes in the activity to define an ordered sequence among them. Four of the nodes are actions: two call behavior actions, one wait time action, and one accept event action (specialized kinds of actions discussed in detail in Section 6.8). When one action completes, it offers a control token on its outgoing control flow, and that enables the next action in the sequence to begin.

6.8 Actions, Revisited

Now that you know about object nodes and edges, let's look at the rest of what you need to know about actions. Let's look at when actions start and discuss four specialized kinds of actions: call behavior actions, send signal actions, accept event actions, and wait time actions.

6.8.1 When Does an Action Start?

To correctly interpret and create activity diagrams, it's essential that you understand when an action will start. Three conditions must be satisfied for an action to start:

- The activity that owns the action is currently executing.
- A control token arrives on *each* of the incoming control flows.
- A sufficient number of object tokens arrive on each of the incoming object flows to satisfy the lower multiplicity of the respective input pin.

Let's look at these rules in the context of a few sample actions, as shown in Figure 6.13.

The action *ma* has one incoming control flow and one incoming object flow. The object flow is attached to an input pin, *sensorSelection*, which has a lower multiplicity of zero. This action therefore starts the moment a control token arrives on the incoming control flow, whether or not an object token also is available on the pin *sensorSelection*.

The action *vc* has one incoming control flow and one incoming object flow. The object flow is attached to an input pin that requires object tokens of type *Transfer Command*. There's no multiplicity displayed for this pin. The multiplicity is therefore 1..1 by default. This means that

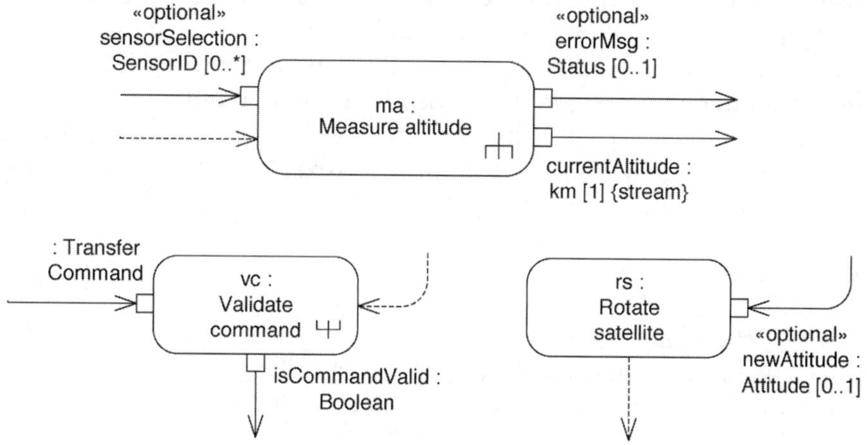

Figure 6.13 *Actions with incoming edges*

the action will start when one control token *and* one object token arrive on their respective incoming edges. These tokens need not arrive at the same time (and likely will not). However, both must be present on the incoming edges for the action to start.

The action *rs* has one incoming object flow, which is attached to an input pin, *newAttitude*. This input pin is optional; it has a lower multiplicity of zero. To start, this action requires no tokens (of either kind). In this case, only the first of the three conditions applies; this action will start the moment its owning activity begins executing.

The most common mistake that modelers make on activity diagrams is drawing multiple incoming edges to an action to convey alternative paths to that action. They mistakenly believe that multiple incoming edges represents an *or* condition. As you now know, it represents an *and* condition. Please be vigilant in looking for this error in your own activity diagrams; it's an easy mistake. If you need to model alternative paths to a given action, you must insert a merge node preceding the action (more on that in Section 6.9.4, "Merge Nodes").

Be aware, also, that an action need not have any incoming edges at all. In that case, you're conveying that the action is not waiting for any input tokens; it will start as soon as its owning activity starts. And if there are *multiple* actions in the activity with no incoming edges, they *all* start concurrently.

6.8.2 Call Behavior Actions

A **call behavior action** is a specialized action that invokes another behavior when it becomes enabled. Call behavior actions let you decompose a higher-level behavior into a set of lower-level behaviors.

A call behavior action becomes enabled according to the same rules just discussed. The behavior that it calls can be any one of the three kinds: an interaction, a state machine, or another activity.

The notation for a call behavior action is the same as the one for an action—a rectangle with rounded corners—except that the name string inside has a particular format:

```
<action name> : <Behavior Name>
```

The action name is modeler defined. The behavior name must match the name of an interaction, a state machine, or an activity that you've defined somewhere in your model hierarchy.

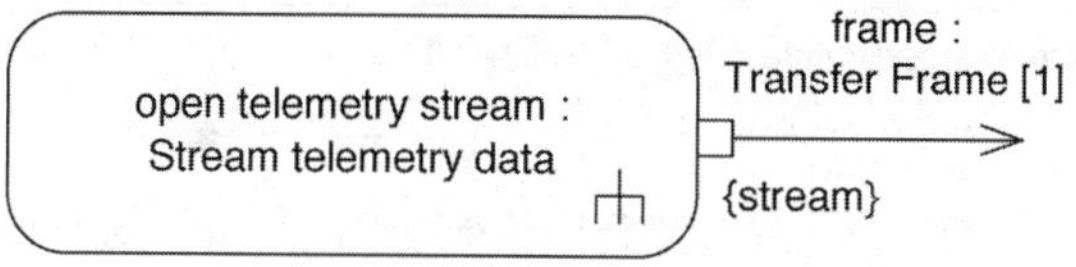

Figure 6.14 *A call behavior action that invokes the behavior* Stream telemetry data

All three of the actions shown in Figure 6.13 are call behavior actions. You can conclude from this that *Measure altitude, Validate command,* and *Rotate satellite* are the names of three behaviors defined within the system model. If a rake symbol (⋔) appears in the lower-right corner of a call behavior action (such as *ma* or *vc*), it conveys that the behavior getting called (*Measure altitude, Validate command*) is an activity. A call behavior action without a rake symbol (such as *rs*) is ambiguous; the called behavior (*Rotate satellite*) can be either an interaction or a state machine. With that said, it's rare for modelers to use call behavior actions to invoke state machines.

Figure 6.14 shows an example of a call behavior action named *open telemetry stream,* which invokes the behavior *Stream telemetry data.* The rake symbol conveys that *Stream telemetry data* is an activity. This call behavior action has one output pin named *frame,* which holds an object token of type *Transfer Frame.* Because this pin is a streaming pin, the object token can emerge from the called activity and become available on the pin—for downstream actions to consume—even while *Stream telemetry data* is executing. This action could therefore produce multiple object tokens of type *Transfer Frame* in the course of a single execution.

When a call behavior action invokes another activity, the pins of the call behavior action must match the activity parameters of the called activity. Figure 6.15 shows the definition of the *Stream telemetry data* activity, which owns an output activity parameter that matches the output pin shown in Figure 6.14. Once invoked, this activity will continually transform object tokens of type *Source Packet* from multiple data sources into a single stream of object tokens of type *Transfer Frame.* As each object token arrives at the output activity parameter, *frame,* it is immediately posted to the corresponding output pin of the call behavior action, *open telemetry stream.*

You can use call behavior actions to factor out a common chunk of functionality that appears in multiple places and instead define it in a separate behavior that you simply invoke multiple times. This design

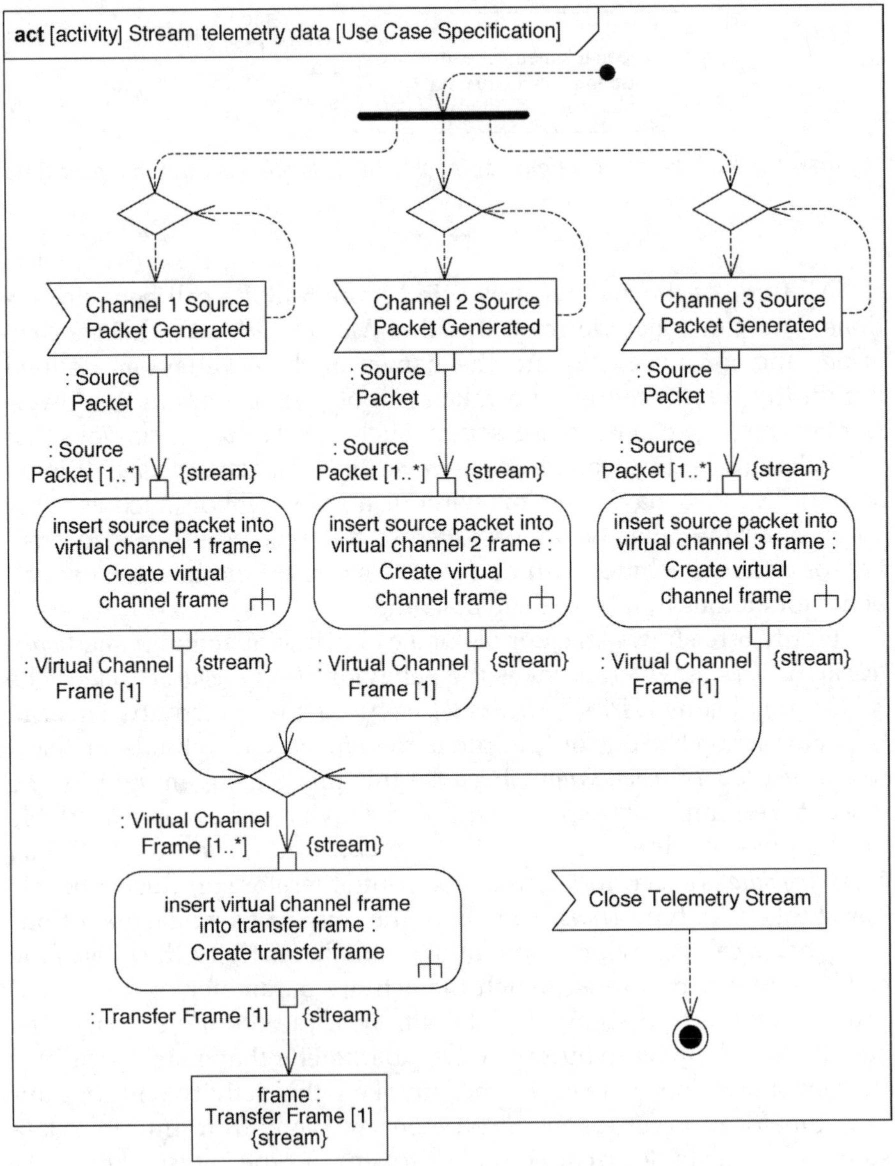

Figure 6.15 *The definition of the* Stream telemetry data *behavior*

practice facilitates easy reuse of common lower-level behaviors. For example, the *Stream telemetry data* activity contains three call behavior actions, all of which invoke the *Create virtual channel frame* behavior. That behavior is defined once—as an activity somewhere in the model hierarchy—and executed here multiple times. (And in this particular case, those three distinct executions occur in parallel.)

6.8.3 Send Signal Actions

To satisfy scalability and performance requirements, systems engineers often design systems that are distributed and concurrent. Such systems use asynchronous mechanisms to transfer matter, energy, or data and to synchronize the actions of parts that operate in parallel. You can model this type of system behavior on activity diagrams using send signal actions and accept event actions.

A **send signal action** is a specialized kind of action that asynchronously generates and sends a signal instance to a target when it becomes enabled. A send signal action becomes enabled according to the same rules discussed in Section 6.8.1.

The notation for a send signal action is a convex pentagon shaped like a signpost (as shown in the lower half of the activity diagram in Figure 6.16). The string displayed inside a send signal action (e.g., *Orbit Radius Updated*) must match the name of a signal that you've defined somewhere in your model hierarchy.

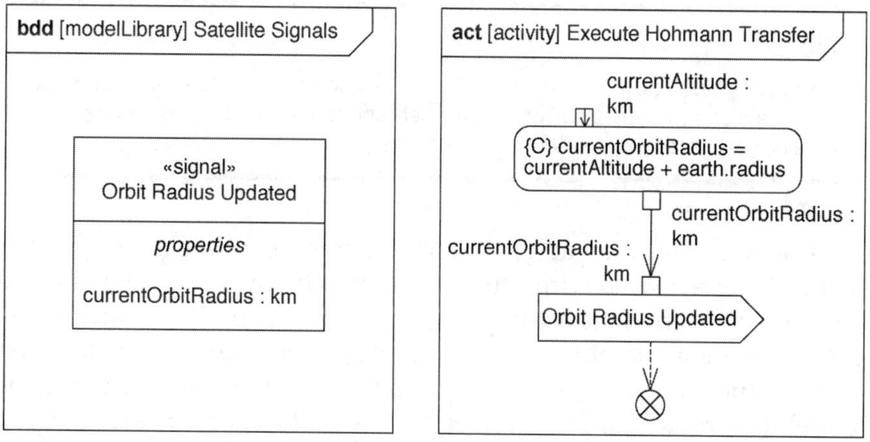

Figure 6.16 *A signal and a send signal action*

Recall from Chapter 3, "Block Definition Diagrams," that a signal is a kind of model element. Like a block, a signal can own properties. Those properties generally represent data that a signal instance carries from a sender to a target. The BDD in Figure 6.16 displays the *Orbit Radius Updated* signal. This signal owns one property, *currentOrbitRadius*, of type *km*. An instance of this signal can therefore carry a single *km* value from a sender to a target.

The activity diagram in Figure 6.16 displays an excerpt of the *Execute Hohmann Transfer* activity shown in Figure 6.1, with a focus on the send signal action. When this send signal action becomes enabled—upon the arrival of an incoming object token at its input pin—it asynchronously generates an instance of the *Orbit Radius Updated* signal, which carries the *currentOrbitRadius* value to a target that's waiting for it. Because the send signal action is asynchronous, it doesn't wait for a response from that target; instead, it completes immediately and offers a control token on its outgoing edge. The send signal action is again enabled each time a new object token arrives at its input pin.

6.8.4 Accept Event Actions

An **accept event action** is the partner of the send signal action in asynchronous behaviors; an accept event action is the element you use in an activity to convey that the activity must wait for an asynchronous event before it can continue its execution. Typically, that asynchronous event occurrence is the receipt of a signal instance.

Note

An accept event action is not limited to receiving signal instances (from send signal actions); it can also receive asynchronous time event occurrences (for details, see Section 6.8.5, "Wait Time Actions").

The notation for an accept event action is a concave pentagon that looks like a rectangle with a triangular notch cut out on one side (as shown center-left in the activity diagram in Figure 6.17). The string displayed inside an accept event action (e.g., *Orbit Radius Updated*) often matches the name of a signal that you've defined somewhere in your model hierarchy. This is how you convey that the accept event action waits for an instance of that signal to arrive asynchronously. When it does, the accept event action completes, and the flow of control proceeds to the next node in the activity.

Note

The accept event action that receives a signal instance may appear in the same activity as the send signal action that generates it. Or it may instead appear in a separate activity; in this way, you can model asynchronous communication between two distinct system behaviors.

Figure 6.17 displays an excerpt of the *Execute Hohmann Transfer* activity shown in Figure 6.1. This excerpt focuses on the asynchronous communication that occurs between two concurrent flows of control through the activity. Concurrent flows proceed independently, but sometimes they must be synchronized at certain points in the behavior.

In this example, the accept event action becomes enabled when a control token arrives on its incoming control flow, something that happens only after the *enter transfer orbit* action completes. If the send signal action on the right side has already generated an *Orbit Radius*

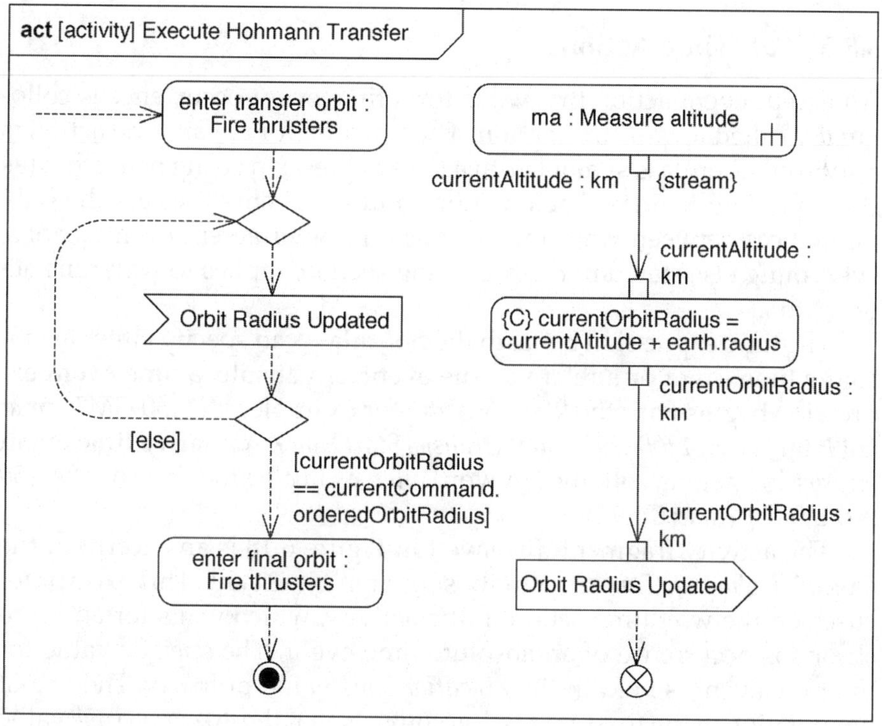

Figure 6.17 *An activity fragment containing an accept event action*

Updated signal instance, then the accept event action completes immediately and offers a control token on its outgoing control flow; then execution proceeds to the next node.

On the other hand, if the send signal action on the right side has not generated an *Orbit Radius Updated* signal instance, then the accept event action simply waits for an *Orbit Radius Updated* signal instance; the *Execute Hohmann Transfer* activity cannot proceed to the *enter final orbit* action until that occurs.

Accept event actions, like other kinds of actions, need not have any incoming edges at all. And the rules in Section 6.8.1 state that an action with no incoming edges becomes enabled the moment the activity begins executing. An accept event action is no exception in this respect; an accept event action with no incoming edges becomes enabled and begins listening for a signal instance as soon as the activity begins executing.

However, there is one difference: An accept event action with no incoming edges remains enabled even after the first signal instance arrives, and the accept event action continues to listen for additional instances. This is one way to model a system behavior that continually responds to asynchronous event occurrences.

6.8.5 Wait Time Actions

An accept event action that waits for a time event occurrence is colloquially called a **wait time action**. The notation for a wait time action is a stylized hourglass symbol with a time expression string beneath it (as shown in Figure 6.18). The notations and the event types are the only differences between wait time actions and other accept event actions; everything else stated in the preceding section applies to wait time actions, too.

The time expression beneath the hourglass can specify either an absolute time event or a relative time event. An **absolute time event** expression begins with the keyword *at*—for example, *at (1430 GMT)* or *at (14 NOV 2106, 1200 CST)* or *at (transferStartTime)*. A **relative time event** expression begins with the keyword *after*, as in *after (30 days)* or *after (50 ms)* or *after (timerCount)*.

The activity fragment displayed in Figure 6.18 is an excerpt of the *Execute Hohmann Transfer* activity shown in Figure 6.1. This excerpt focuses on the wait time action in the activity, which waits for an asynchronous occurrence of an absolute time event. The specific value for the time event is held in the *executionTime* value property. The use of dot notation in the time expression conveys that the *executionTime* value property is nested within the *currentCommand* object.

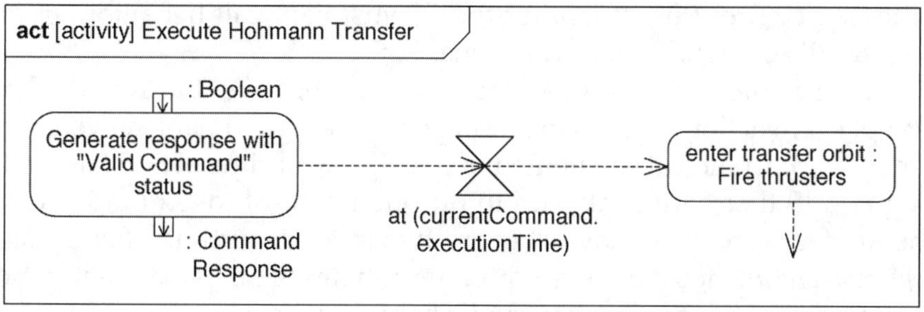

Figure 6.18 *A wait time action with an absolute time event expression*

This wait time action becomes enabled when a control token arrives on its incoming control flow—upon completion of the upstream action that generates a valid command response. If the absolute time event—specified by *executionTime*—has already occurred, then the wait time action completes immediately and offers a control token on its outgoing control flow; then execution proceeds to the next node. If *executionTime* has not occurred, then the wait time action simply waits for that time event to occur; the *Execute Hohmann Transfer* activity cannot proceed to the *enter transfer orbit* action before that moment occurs.

When a wait time action has a relative time event expression—as shown in Figure 6.19—the clock for the time event begins when the wait time action becomes enabled. Because the wait time action shown has no incoming edges, it becomes enabled as soon as the *Sample system temperature* activity begins executing. Two seconds later, the relative time event occurs, and the wait time action offers a control token on its

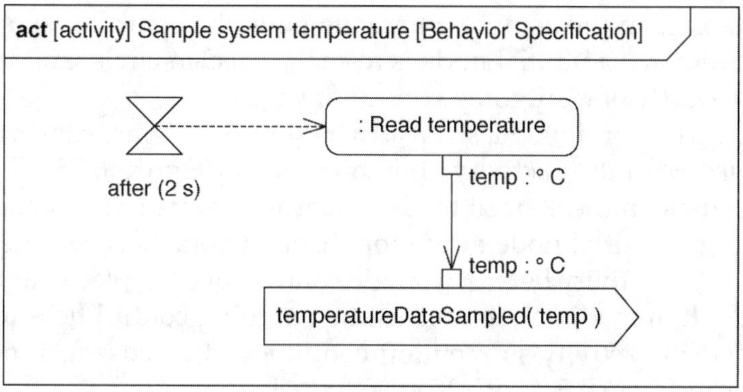

Figure 6.19 *A wait time action with a relative time event expression*

outgoing control flow, enabling the downstream call behavior action, which calls the *Read temperature* behavior.

A wait time action is a special kind of accept event action. And an accept event action with no incoming edges remains enabled even after the first occurrence of the named event. The wait time action shown in Figure 6.19 therefore continues to output a control token (and invoke the *Read temperature* behavior) every two seconds until the *Sample system temperature* activity as a whole terminates. This is an example of how to model a continuous periodic behavior.

6.9 Control Nodes

In this chapter I've discussed two kinds of nodes that can exist within an activity: an action node and an object node. Now let's look at one final kind of node: a control node. You use **control nodes** to steer the execution of an activity along paths other than a simple sequence of actions. Control nodes can direct the flow of control tokens as well as object tokens within an activity.

There are seven kinds of control nodes: initial nodes, activity final nodes, flow final nodes, decision nodes, merge nodes, fork nodes, and join nodes. You can, of course, use combinations of these nodes to define arbitrarily complex control logic in your activities as necessary to satisfy your system's functional requirements.

Let's look at how each one works.

6.9.1 Initial Nodes

An **initial node** marks the starting point within an activity. Formally, it marks a place in the activity where the flow of a control token begins. The notation for an initial node is a small, filled-in circle, and it generally has exactly one outgoing control flow.

The activity fragment shown in Figure 6.20 is an excerpt of the *Stream telemetry data* activity shown earlier in Figure 6.15. This fragment focuses on the start of three concurrent sequences of actions, beginning at an initial node at the top of the diagram. When the *Stream telemetry data* activity begins, a single control token is placed at this initial node. It immediately traverses the outgoing control flow to a fork node, and the activity's execution continues. (I discuss fork nodes in detail in Section 6.9.5, "Fork Nodes.")

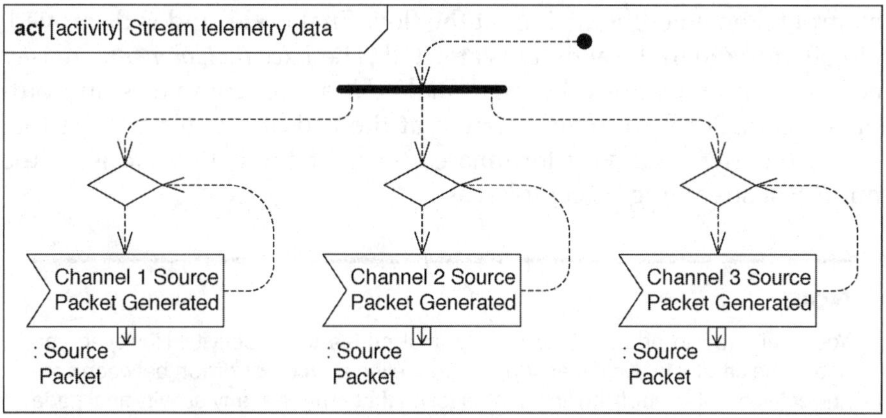

Figure 6.20 *An activity fragment containing an initial node*

Note that an activity need not have an initial node. The flow of a control token can begin instead at an action with no incoming edges. Recall that such actions start as soon as the activity begins executing.

It's also possible that the flow of object tokens alone will be sufficient to define the correct sequences of actions within an activity. Object tokens generally begin at input activity parameters (on the frame of an activity diagram). In such cases, you don't need an initial node in your activity.

6.9.2 Flow Final Nodes and Activity Final Nodes

Flow final nodes and **activity final nodes** are control nodes that mark the end of the flow of a control token. However, there's a key difference between them: When a control token arrives at a flow final node, that token is destroyed, marking the end of a single flow of control. Alternatively, when a control token arrives at an activity final node, the entire activity terminates, marking the end of *all* flows of control (no matter where they are currently in their execution).

The notation for a flow final node is a circle containing an X. The notation for an activity final node is a circle that contains a smaller, filled-in circle. Figure 6.17 (presented earlier) shows an example of each one.

The two sequences of actions shown in Figure 6.17 execute in parallel. The *ma* call behavior action continually streams object tokens on its output pin. The two actions downstream from *ma* may therefore execute multiple times. And each time the send signal action completes, a control token emerges, arrives at the flow final node, and is destroyed. No other token in the activity is affected. The *Execute Hohmann Transfer* activity continues until the *enter final orbit* action completes and outputs a control token, which arrives at the activity final node. At that point, the entire activity terminates (even if the action *ma* is in the middle of streaming object tokens).

Note

You're allowed to add multiple activity final nodes to an activity. However, you should be careful to avoid inadvertently modeling a **race condition** between concurrent flows of control; the first control token that arrives at any activity final node will terminate the entire activity. You should use flow final nodes instead of activity final nodes to avoid modeling a race condition.

6.9.3 Decision Nodes

A **decision node** marks the start of alternative sequences in an activity. The notation is a hollow diamond (as shown in Figure 6.21). A decision node must have a single incoming edge and generally has two or more outgoing edges. Each outgoing edge is labeled with a Boolean expres-

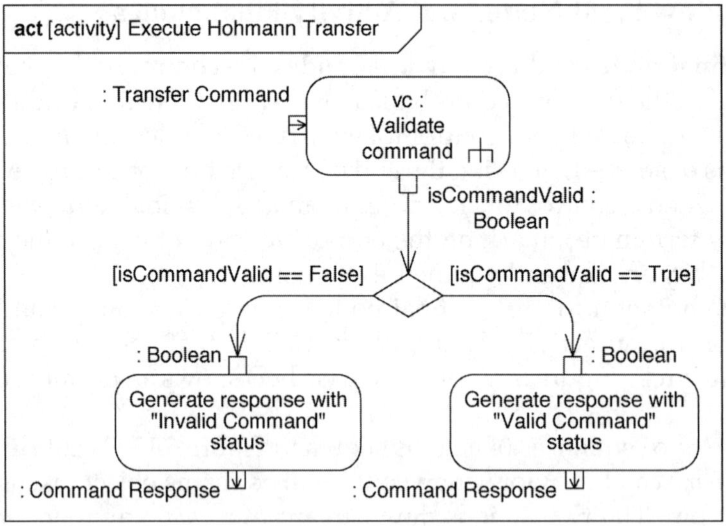

Figure 6.21 *An activity fragment containing a decision node*

sion called a **guard**, which is displayed as a string between square brackets (e.g., *isCommandValid = = False* in Figure 6.21).

When a token—either an object token or a control token—arrives at a decision node, the guards on the outgoing edges are evaluated. The token is offered to the outgoing edge whose guard evaluates to true at that moment.

The onus is on you to ensure that the set of guards on the outgoing edges is complete and disjoint so that exactly one guard evaluates to true each time a token arrives. SysML allows you to use *else* as a guard on (at most) one outgoing edge to ensure the "complete" criterion is met. (If all the other guards evaluate to false, the *else* evaluates to true.)

6.9.4 Merge Nodes

A **merge node** marks the end of alternative sequences in an activity. The notation is the same as the one for a decision node: a hollow diamond. You can distinguish one from the other by the number of incoming and outgoing edges; a merge node has two or more incoming edges and a single outgoing edge. When a token—either an object token or a control token—arrives at a merge node via any of its incoming edges, the token is immediately offered to the outgoing edge.

Most often, you will use a merge node in conjunction with a decision node to model a **loop** within an activity (as shown in Figure 6.22). A merge node is, in fact, essential for modeling a loop. If the merge node in this figure were removed (and its two incoming edges were connected directly to the accept event action), then the accept event action would never become enabled. To begin, it would need a control token on *each* of those incoming edges, and that would never happen in a loop.

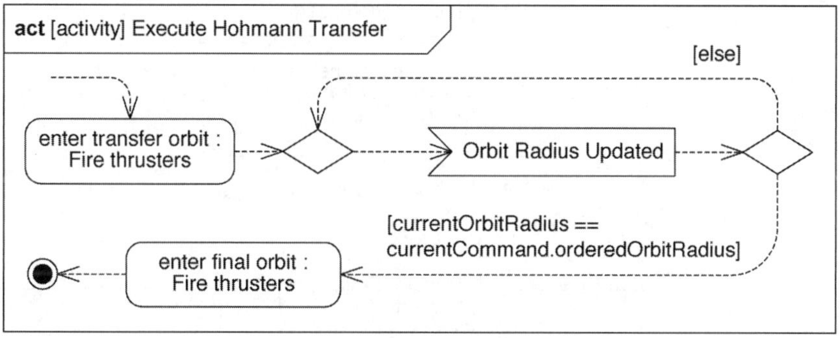

Figure 6.22 *Using a merge node to model a loop*

You can also use a merge node to model the interleaving of tokens from multiple concurrent sources into a single output stream (as shown in Figure 6.23). The three call behavior actions that invoke the *Create virtual channel frame* activity execute concurrently in this activity. They independently output object tokens of type *Virtual Channel Frame*, which arrive at the merge node in a nondeterministic order. As each one arrives, it is immediately offered to the outgoing edge and becomes an input to the next action. That action, in turn, outputs a single stream of object tokens of type *Transfer Frame*.

6.9.5 Fork Nodes

A **fork node** marks the start of concurrent sequences in an activity. The notation for a fork node is a line segment (oriented in any direction you like), and it must have a single incoming edge and two or more outgo-

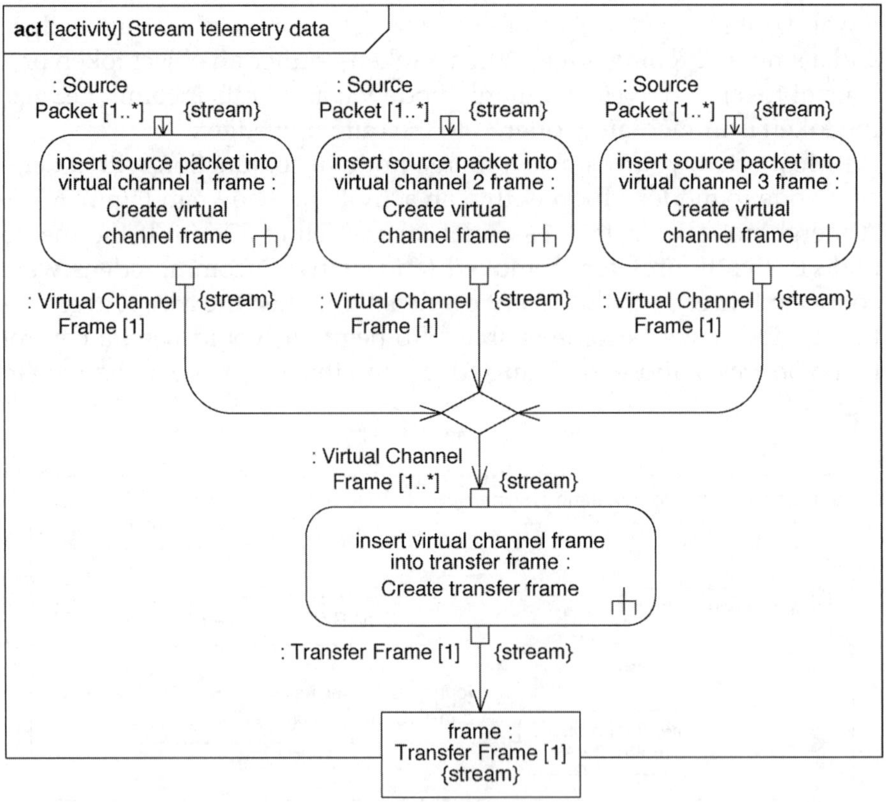

Figure 6.23 *Using a merge node to model the interleaving of tokens*

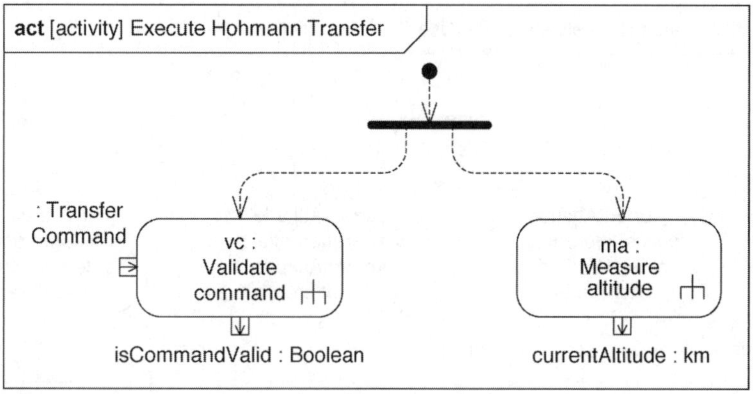

Figure 6.24 *An activity fragment containing a fork node*

ing edges (as shown in Figure 6.24). When a token—either an object token or a control token—arrives at a fork node, it is duplicated on *all* of the outgoing edges. Each of those copies of the original token represents an independent, concurrent flow of control along its respective path.

The activity fragment shown in Figure 6.24 is an excerpt of the *Execute Hohmann Transfer* activity shown earlier in Figure 6.1. This excerpt focuses on the fork node in the activity. When a control token arrives at the fork node, it gets duplicated on the two outgoing edges. Those two copies then proceed to their respective downstream actions: *vc* and *ma*. The action *ma* begins immediately; it's not waiting on any other input. The action *vc* begins as soon as an object token of type *Transfer Command* arrives on its input pin. Depending on when that occurs, the two actions may execute concurrently for some period.

An important general point about concurrency is that the order of completion of concurrent actions is nondeterministic. You cannot know ahead of time whether *vc* or *ma* will complete first in a given execution of the *Execute Hohmann Transfer* activity during system operation. More precisely, it doesn't matter which one completes first; you should model actions as concurrent only when they have no dependencies on one another.

6.9.6 Join Nodes

A **join node** marks the end of concurrent sequences in an activity. The notation for a join node is the same as the one for a fork node: a line

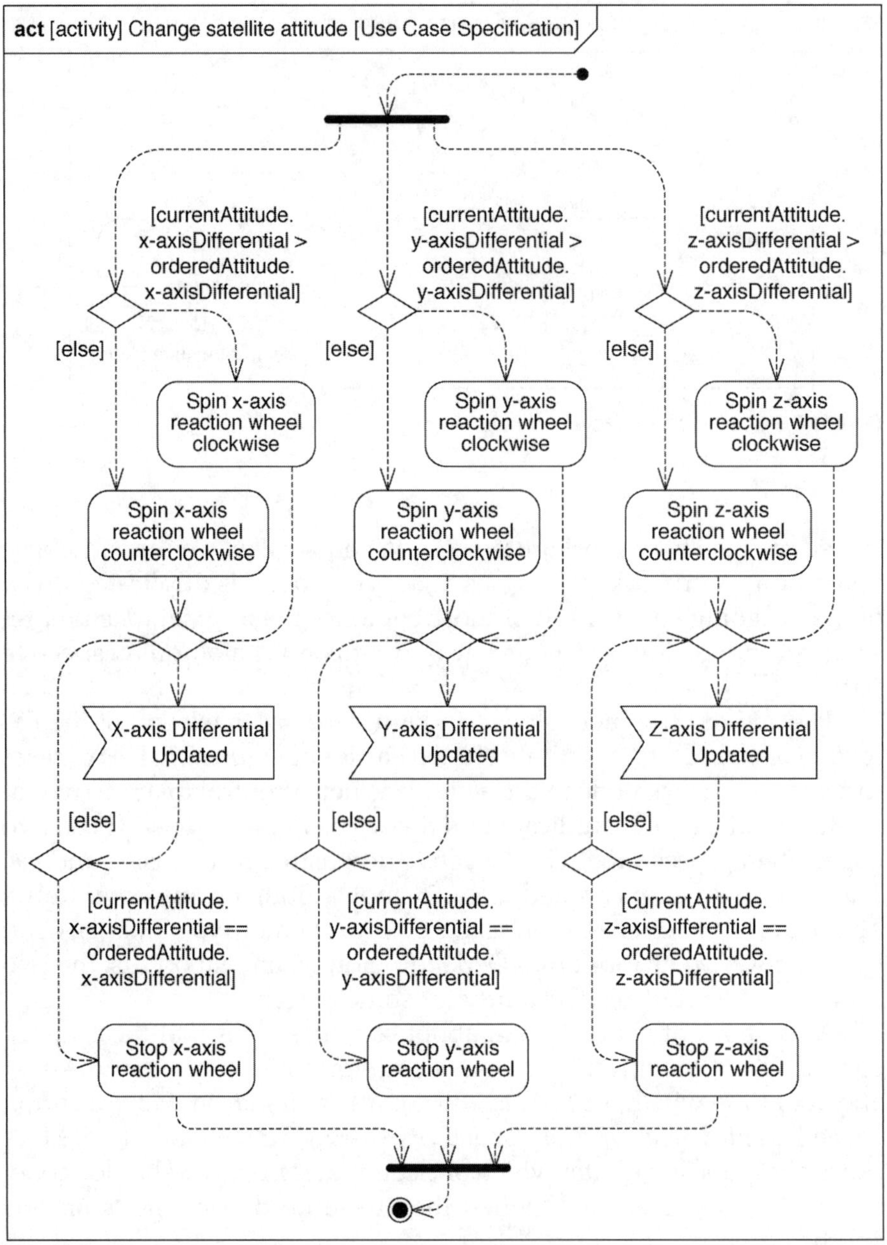

Figure 6.25 *Using a join node to synchronize the end of concurrent sequences of actions*

segment. You can distinguish one from the other by the number of incoming and outgoing edges; a join node generally has two or more incoming edges and a single outgoing edge (as shown at the bottom of the activity diagram in Figure 6.25).

You use a join node to model a synchronization point for concurrent sequences of actions in an activity. When a token arrives on *each* of the incoming edges, a single token is offered on the outgoing edge. The concurrent sequences end, and a single flow of control proceeds past the point marked by the join node.

The activity shown in Figure 6.25 has three concurrent sequences of actions, which begin at the fork node at the top of the diagram. These sequences of actions model the starting and stopping of three reaction wheels, each of which rotates the satellite about one axis independently of the other two. The presence of the join node at the bottom of the diagram conveys that all three sequences of actions must complete before the activity can come to an end.

6.10 Activity Partitions: Allocating Behaviors to Structures

Activity diagrams can convey not only the order of the actions within an activity but also the structure that performs each action. You use **activity partitions** to convey this. The notation for an activity partition is a large rectangle (which contains one or more nodes) with a header at one end; the header specifies what the activity partition represents. An activity partition may be oriented either vertically or horizontally, although some modeling tools may support only one direction or the other.

Most often, an activity partition represents either a block or a part property that exists somewhere in the system model. Placing an action inside an activity partition conveys that the action is allocated to the structure named in the header. Simply put, that structure is responsible for performing that action when the activity executes during system operation.

When an activity partition represents a block, it conveys that *all* instances of that block can perform the contained actions. When an activity partition represents a part property, only that one part property is responsible for performing the contained actions.

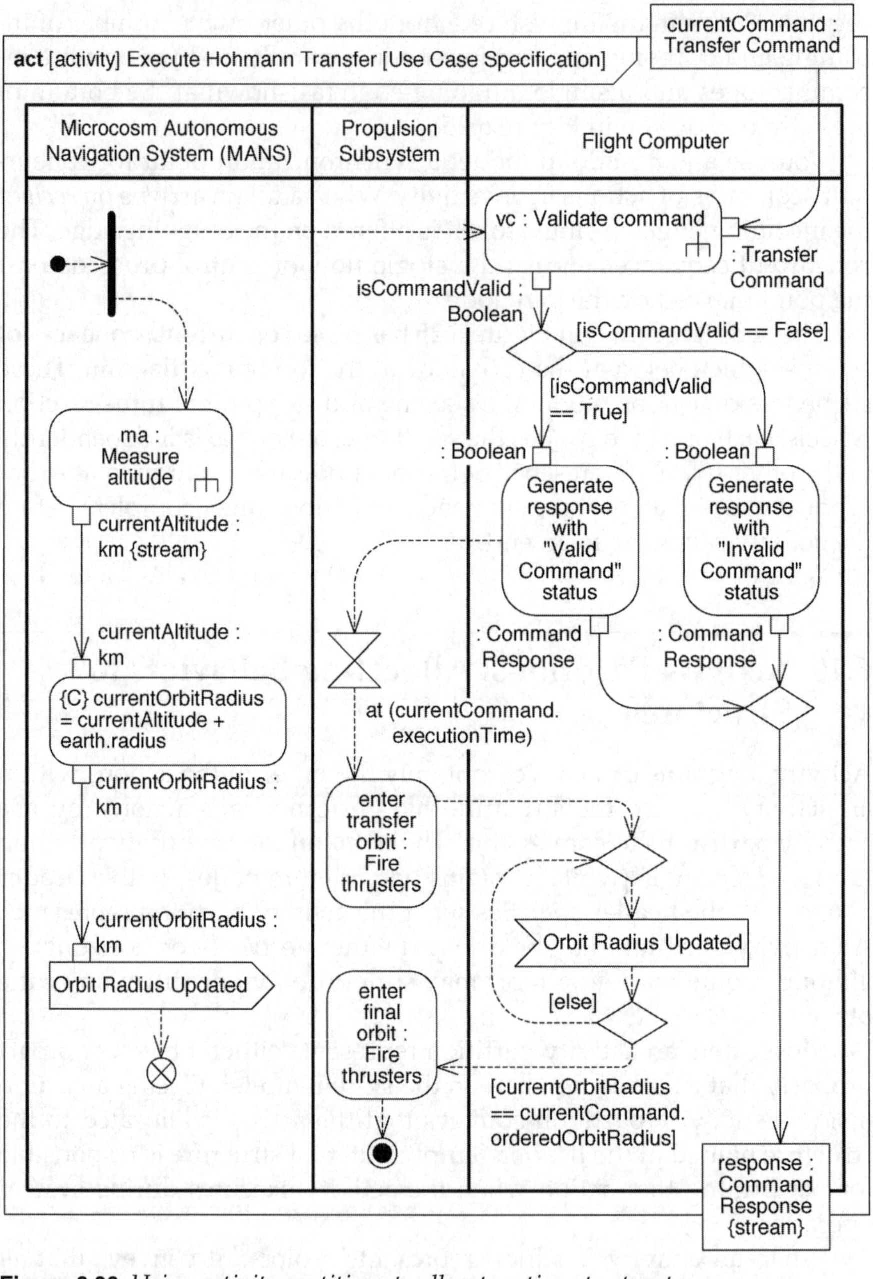

Figure 6.26 *Using activity partitions to allocate actions to structures*

Figure 6.26 provides an alternative view of the complete *Execute Hohmann Transfer* activity shown in Figure 6.1. All of the same actions as before are displayed here. However, this view also displays three activity partitions, which convey the allocations of those actions to three different blocks in the system model: *Microcosm Autonomous Navigation System (MANS)*, *Propulsion Subsystem*, and *Flight Computer*. This view conveys a more comprehensive narrative of the *Execute Hohmann Transfer* behavior—a narrative that links behaviors to structures.

Summary

An activity diagram is a powerful medium for communicating information to your stakeholders about a system's behavior over time. This diagram is a good choice when you need to display a continuous system behavior and when you need to put the focus on the flow of matter, energy, and data among a set of actions, whether sequential or concurrent. An important strength of an activity diagram is its readability, even when displaying a behavior having complex control logic.

You can create call behavior actions in an activity to model behavioral decomposition. Send signal actions and accept event actions enable you to model asynchronous communication among the structures within a distributed system. You can use wait time actions to model behaviors that occur periodically or begin at particular moments in time. Activity partitions let you allocate responsibility for the actions in an activity to specific structures within a system. All these features make an activity diagram a richly expressive medium for conveying system behavior.

Chapter 7

Sequence Diagrams

Sequence diagrams are a second kind of SysML diagram that you can use to express information about a system's dynamic behavior. You can use elements called lifelines to model the participants in a system behavior and then use messages between lifelines to model interactions among those participants. You can specify time constraints and duration constraints on interactions. You can also use various kinds of interaction operators to steer the execution of an interaction. Interaction uses let you model behavioral decomposition among a set of interactions.

7.1 Purpose

A **sequence diagram** is a kind of behavior diagram; like an activity diagram, it presents a dynamic view of the system, a view that expresses sequences of behaviors and event occurrences over time. A sequence diagram is a good choice for specifying a behavior when you want the focus to be on how the parts of a block interact with one another via operation calls and asynchronous signals to produce an emergent behavior. That behavior is formally called an **interaction**.

In Chapter 6, "Activity Diagrams," you learned that SysML gives you three kinds of behavior diagrams for specifying a behavior: sequence diagrams, activity diagrams, and state machine diagrams. Each one has strengths and weaknesses that you should be aware of to make the best choice based on the needs of your target audience.

A sequence diagram is a precise specification of a behavior. It is therefore well suited to serve as a detailed design artifact—an input into development. Many commercial-grade modeling tools blur the line between design and development; they let you autogenerate production-quality source code based on the interactions displayed on sequence diagrams. This is possible because sequence diagrams convey the three requisite pieces of information needed to automate a transformation into source code: the order in which behaviors are performed, which structure performs each behavior, and which structure invokes each behavior.

That precision, however, comes at the cost of readability. A sequence diagram quickly becomes unreadable as the complexity of the control logic in the behavior increases. For this reason, modelers often use a sequence diagram as a graphical test case specification that displays only a single scenario of a larger (more complex) use case behavior. However, it's not limited to this conventional use. A sequence diagram is a sufficiently rich medium to serve as a general-purpose behavior specification tool.

7.2 When Should You Create a Sequence Diagram?

Because a sequence diagram is a versatile behavior diagram, you can create one to specify a behavior at any level in the system hierarchy. Sequence diagrams are useful early in the life cycle during ConOps development to specify the intended interactions between the system of interest and the actors in its environment. These diagrams are useful early in the architectural design stage to specify the interactions between subsystems. And they're useful at the end of architectural design to specify the interactions between components in preparation for the hand-off to the teams producing detailed designs of components.

If your team chooses to use sequence diagrams as graphical test case specifications, you would also create sequence diagrams in place of the traditional text-based test case specifications. The exact point in the life cycle when this occurs varies greatly from one project team to the next. Teams that practice test-driven design and development, for example, create their test case specifications in parallel with the requirements specification (or use case specification) activity.

Sequence diagrams are a good choice whenever you need to precisely specify the interactions between entities, within either the problem space or the solution space for your system of interest. In short, you can potentially create a sequence diagram at any point in the system life cycle.

7.3 The Sequence Diagram Frame

The diagram kind abbreviation for a sequence diagram is *sd*. The only allowable model element type for a sequence diagram is *interaction*. The frame of a sequence diagram always represents a single interaction that you've defined somewhere in your system model.

An interaction is itself a model element; like an activity, it's a kind of behavior. Like an activity, a block, and a package, an interaction is also a kind of namespace. It can therefore contain a set of named elements (such as lifelines, event occurrences, and messages) within the model hierarchy. Those contained elements can appear within the frame of the associated sequence diagram.

The diagram header in Figure 7.1 tells us that the frame of this sequence diagram represents the interaction named *Execute Hohmann Transfer, Main Success Scenario*, which is defined somewhere in the system model. The elements that appear within the frame are contained within (nested under) this interaction within the model hierarchy.

Figure 6.26 (in Chapter 6) displays the complete *Execute Hohmann Transfer* use case specification. The sequence diagram in Figure 7.1 displays the main success scenario within that use case specification—the behavior that occurs when the satellite receives a valid transfer command. Because these two figures show substantially similar behaviors, they provide a fair comparison of the readability of sequence diagrams compared with activity diagrams.

As mentioned earlier in this chapter, the precision of a sequence diagram comes at the cost of readability; the sequence diagram in Figure 7.1 is quite cluttered. In this chapter, I analyze this diagram to explain in detail each kind of element that can appear on a sequence diagram.

7.4 Lifelines

A **lifeline** is an element that represents a participant in an interaction (see Figure 7.2). More precisely, a lifeline represents a single instance that participates in an interaction by exchanging messages with other

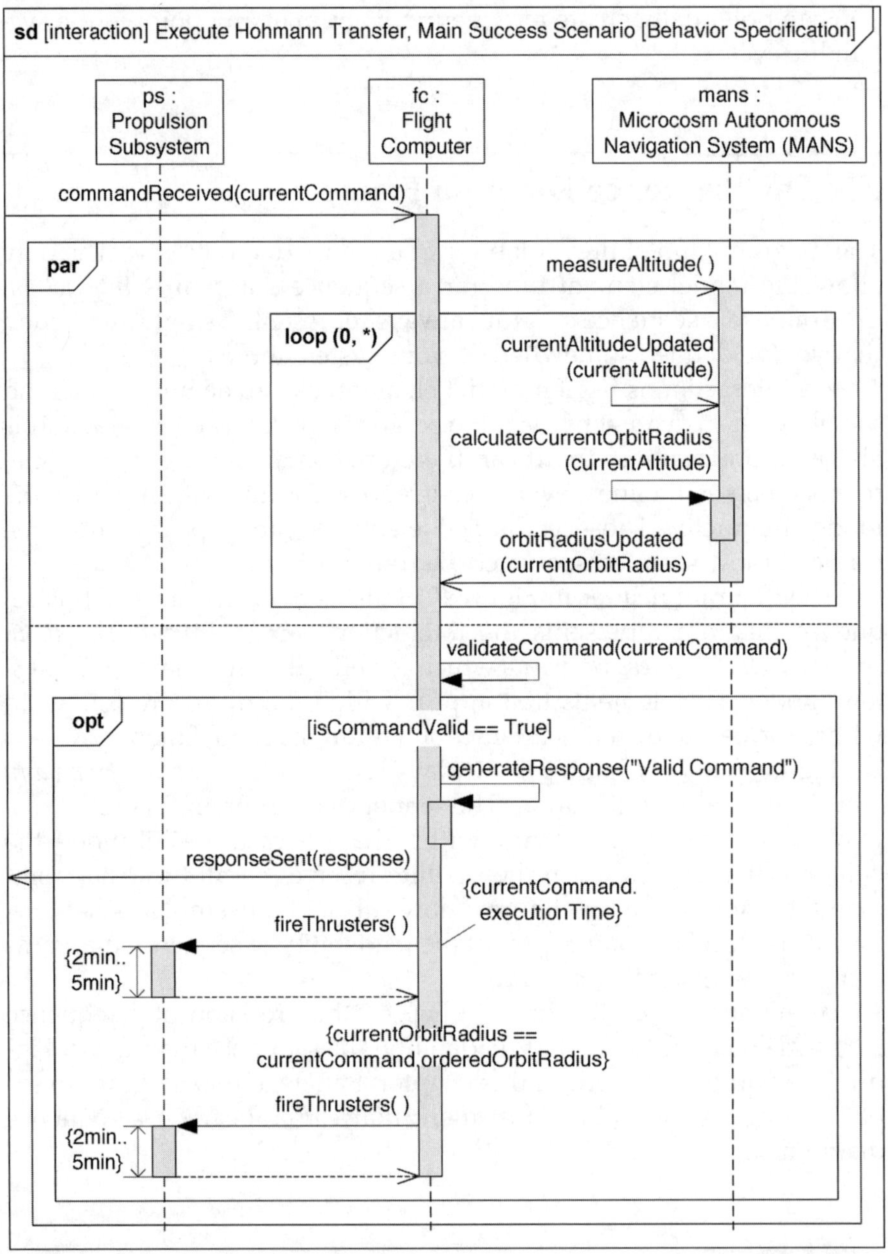

Figure 7.1 *A sample sequence diagram*

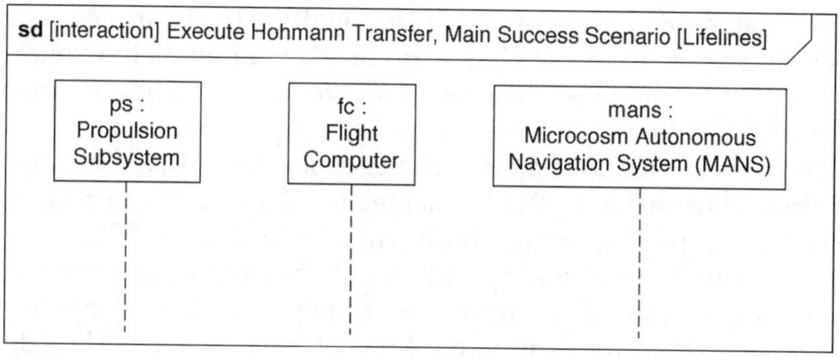

Figure 7.2 *Lifelines*

lifelines. The lifelines that appear in a given interaction correspond to part properties of the block that owns the interaction. That block may be the system of interest as a whole, a subsystem, or a single component.

The notation for a lifeline is a rectangle with a dashed line attached to it, flowing down the sequence diagram. The dashed line represents the lifetime of the part property (with respect to its participation in the interaction, and not necessarily its entire existence in an operational system). Time proceeds down the lifeline; an event occurrence that appears higher on a lifeline happens before one that appears lower on that same lifeline.

A sequence diagram, however, conveys only the *relative* passage of time between event occurrences. The linear distance between two event occurrences on a lifeline means nothing; all that matters is which one is higher and which one is lower. (You can read more on event occurrences later in this section.)

The rectangle is referred to as the **head** of the lifeline. It contains a name string that identifies the part property that the lifeline represents. That name string appears in the following format:

```
<part property name> [<selector expression>] : <type>
```

The type of the named part property is a block or an actor you've defined somewhere in your model hierarchy. Most often, it is a block. However, if your team chooses to use sequence diagrams to display the interactions at the domain level (between the system of interest and the actors in its environment), then you will also have lifelines typed by actors.

The **selector expression** is an optional part of the name string. When displayed, it appears in square brackets immediately following the name of the lifeline. The selector expression specifies a particular instance the lifeline represents.

This concept requires a bit of elaboration. Recall that a lifeline represents a single instance, and the name of a lifeline must correspond to the name of a part property. You learned in Chapter 3, "Block Definition Diagrams," that a single part property may represent an entire collection of instances (if the upper multiplicity of the part property is greater than 1). With that in mind, I repeat the following: The selector expression specifies a particular instance (in a collection) that the lifeline represents.

The BDD at the right in Figure 7.3 shows that the *Software Subsystem* block owns a part property named *ssdd* of type *Star Sensor Device Driver*. This part property has a multiplicity of 3, which conveys that it represents exactly three instances of *Star Sensor Device Driver*; the *ssdd* part property therefore represents a collection of instances. The sequence diagram in Figure 7.3 has a lifeline named *ssdd*, which corresponds to the part property of the same name. This lifeline specifies a selector expression, *x-axis*, to name the particular instance in the *ssdd* collection that the lifeline represents in this interaction.

As mentioned earlier, however, a selector expression is optional; you need to add one to a lifeline only when it actually matters which instance in the collection participates in the interaction. If you omit the selector expression, the lifeline represents an arbitrarily chosen instance; you're conveying that it doesn't matter which instance is the one that actually participates in the interaction.

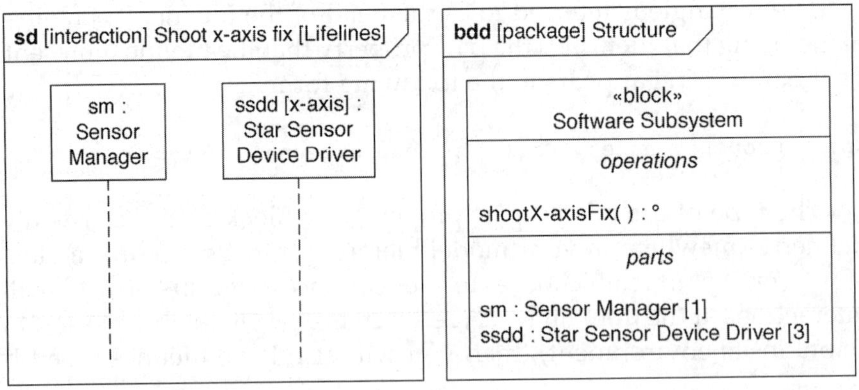

Figure 7.3 *A lifeline with a selector expression*

Earlier in this section, I state that a lifeline conveys the passage of time (going down the diagram). I then reference the concept of event occurrences, which appear on lifelines. The key idea is this: The set of lifelines in an interaction convey ordered sequences of event occurrences, and those sequences form the narrative of the interaction.

There are six kinds of event occurrences that can appear on lifelines:

- Message send occurrences
- Message receive occurrences
- Lifeline creation occurrences
- Lifeline destruction occurrences
- Behavior execution start occurrences
- Behavior execution termination occurrences

In the following sections, I discuss how to represent each of these kinds of event occurrences on a sequence diagram to convey a meaningful narrative.

7.5 Messages

A **message** represents a communication between a sending lifeline and a receiving lifeline. That communication may be an invocation of a behavior, a reply at the end of a behavior, the creation of a lifeline, or the destruction of a lifeline. I discuss these various types of messages in detail in Section 7.5.2, "Message Types."

The notation for a message generally is a line with an arrowhead, connecting the sending and receiving lifelines. The tail end of the message is always connected to the sending lifeline; the arrowhead end of the message is always connected to the receiving lifeline. Each type of message has a distinctive line style (such as dashed or solid) and arrowhead style (such as open or filled in).

A message also has a name string floating above it that specifies the name of the message, along with other optional pieces of information (such as parameter names, argument values, and return value). The exact format of the name string depends on the message type. I provide the formats when I discuss each message type in Section 7.5.2.

Most often, the sending lifeline and the receiving lifeline are two distinct lifelines. However, it's entirely permissible for one lifeline to be both the sender and the receiver of a message. Four examples of this are shown in Figure 7.1. This conveys either that the lifeline sends a communication to itself (for example, to invoke an internal behavior) or that the lifeline is a black-box representation of an entity composed of internal parts that are communicating with each other.

When systems engineers use the term *message*, it often connotes some kind of data format. Please do not carry this predisposition into your study of SysML. In the context of interactions, the term *message* has no such connotation. A message between two lifelines could represent a call to an operation, for example, even if no data are passed in with that message. You can, of course, convey to your readers that data (or objects generally) *are* passed between lifelines (by displaying arguments in the name string for a message). However, the term *message* never implies that.

7.5.1 Message Occurrences

There are six kinds of event occurrences that can appear on lifelines in an interaction. Two of those kinds of event occurrences are message send occurrences and message receive occurrences. We can refer to either of these kinds more generally as **message occurrences**.

Every time you create a message from one lifeline to another (or from one lifeline back to itself), you are modeling both a message send occurrence and a message receive occurrence. A **message send** occurrence exists on a lifeline at the point where the tail end of a message meets the lifeline. A **message receive** occurrence exists on a lifeline at the point where the arrowhead end of a message meets the lifeline. To be clear, there's no notation for a message occurrence; one is implicitly there at the point where a message end connects to a lifeline.

The sequence diagram in Figure 7.4 is an excerpt of the larger interaction shown in Figure 7.1. This excerpt displays a lifeline named *fc*

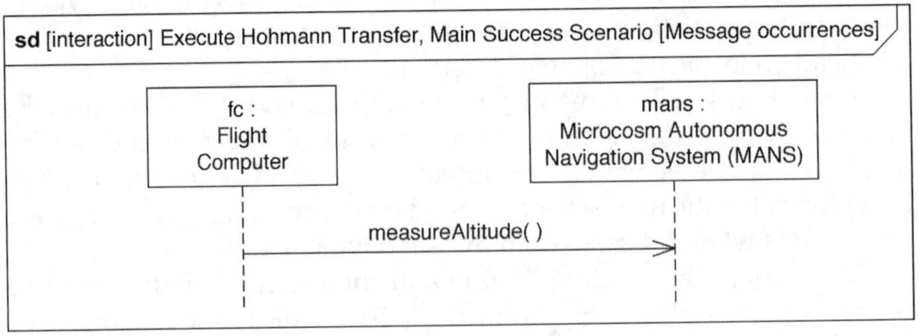

Figure 7.4 *A message send occurrence and a message receive occurrence*

sending a *measureAltitude* message to a lifeline named *mans*. A message send occurrence exists on the *fc* lifeline at the point of intersection with the tail end of the *measureAltitude* message. Similarly, a message receive occurrence exists on the *mans* lifeline at the point of intersection with the arrowhead end of the *measureAltitude* message.

The *measureAltitude* message in Figure 7.4 happens to be an asynchronous message (discussed in more detail in the next section). However, all the key ideas here about message occurrences remain true for the other message types as well.

7.5.2 Message Types

There are four types of messages that commonly appear in interactions: asynchronous messages, synchronous messages, reply messages, and create messages. (SysML defines two other message types—found messages and lost messages—but you'll rarely use these in daily practice.) Each type of message has a distinctive notation. And each type serves a unique purpose in the context of the larger interaction. Let's look at each one in detail.

7.5.2.1 Asynchronous Messages

An **asynchronous message** represents a communication between a sending lifeline and a receiving lifeline wherein the sender immediately proceeds with its own execution after sending the message. The sender does not wait for the receiver to finish executing the invoked behavior, and it does not wait for the receiver to send a reply upon completion of the behavior.

The notation for an asynchronous message is a solid line with an open arrowhead (drawn from the sending lifeline to the receiving lifeline) and a label floating above the line with the following format:

```
<message name> ( <input argument list> )
```

The message name must match the name of a reception owned by the receiving lifeline. (Recall from Chapter 3 that a reception is a kind of behavioral feature that a block can own—a behavioral feature that is always invoked asynchronously.)

The input argument list is an optional piece of information. If you choose not to display it, you can show an empty set of parentheses after the message name—or nothing at all. If you choose to display arguments, they appear in a comma-separated list, and each argument in the list has the following format:

```
<input parameter name> = <value specification>
```

The input parameter name is optional; if shown, it must match the name of an input parameter of the reception that's being invoked. Most often, modelers omit this information and simply display a value specification, which can be either a literal value or the name of a property that holds a value. And, of course, the value that's passed must match the type of its corresponding input parameter (for example, *Integer, Real, Boolean, String, Complex,* or a custom value type or block that you've defined somewhere in your model).

The sequence diagram in Figure 7.5 shows an excerpt of the interaction displayed in Figure 7.1, with an additional message on the *fc* lifeline that doesn't appear in the original interaction. The interaction fragment shown here contains three asynchronous messages: *measureAltitude, currentAltitudeUpdated,* and *orbitRadiusUpdated*. The *measureAltitude* message and the *currentAltitudeUpdated* message each correspond to a reception owned by the *Microcosm Autonomous Navigation System (MANS)* block. Similarly, the *orbitRadiusUpdated* message corresponds to a reception owned by the *Flight Computer* block.

The *measureAltitude* message has no argument list displayed (either because the *measureAltitude* reception has no input parameters or because the arguments for this message are not an important piece of information for this diagram's target audience). The *currentAltitude-*

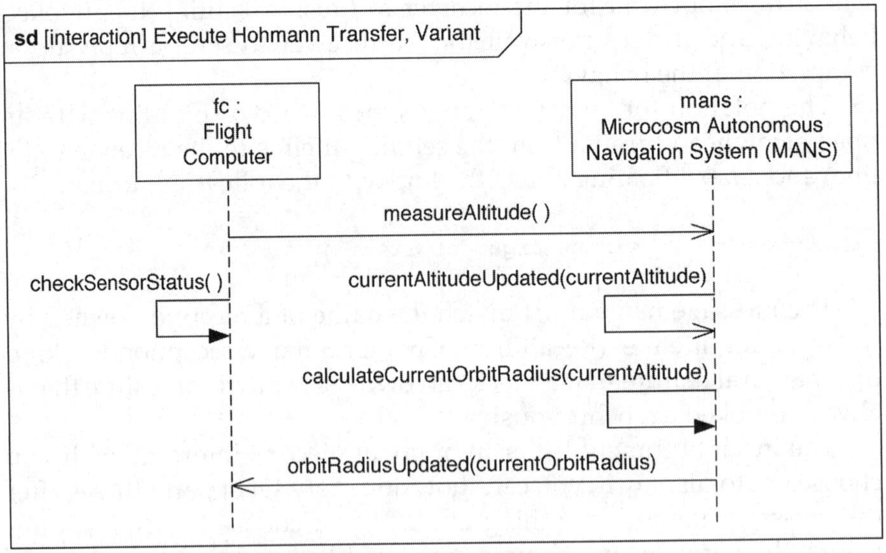

Figure 7.5 *Asynchronous messages in an interaction*

Updated message, however, does have an argument displayed: a property named *currentAltitude*. The value held by this property is the input to the reception; that value can be accessed by the receiving lifeline when it executes the behavior associated with the *currentAltitude-Updated* reception.

The *orbitRadiusUpdated* message, similarly, displays an argument: a property named *currentOrbitRadius*. The *mans* lifeline passes this value in the message to the *fc* lifeline. The *fc* lifeline can therefore access that value when it executes the behavior associated with the *orbitRadius-Updated* message.

Earlier in the chapter, I mention that modelers often create sequence diagrams to serve as graphical test case specifications. When a sequence diagram serves in this role, it's appropriate for the message arguments to be literal values (e.g., *35,786*) instead of property names (e.g., *current-Altitude*). The set of input values throughout the interaction defines a specific test case and enables you to determine (and specify) the set of expected output values. You can then compare that sequence diagram to the results of an actual test execution (or simulation) to deliver a verdict (*pass* or *fail*).

The key idea to take away: The sender of an asynchronous message doesn't wait for the receiver to finish executing the invoked behavior; instead, the sender immediately proceeds with its own execution. In Figure 7.5, for example, *fc* sends an asynchronous message, *measureAltitude*, to *mans*. Upon receipt of that message (or at some nondeterministic time thereafter), *mans* begins executing the behavior associated with the *measureAltitude* reception. The *fc* lifeline, however, does not wait for that behavior to complete; upon sending the *measureAltitude* message, *fc* immediately proceeds with its own execution and sends another message, *checkSensorStatus*.

If *measureAltitude* were instead a synchronous message, the *fc* lifeline would have to wait for *mans* to finish executing the *measureAltitude* behavior and send a reply back to *fc* before *fc* could then proceed to send the *checkSensorStatus* message. And with that, it's time to examine synchronous messages in more detail.

7.5.2.2 *Synchronous Messages*

A **synchronous message** represents a communication between a sending lifeline and a receiving lifeline wherein the sender waits for the receiver to finish executing the invoked behavior and send a reply message before the sender can proceed with its own execution. The notation for a synchronous message is a solid line with a filled-in arrowhead (drawn from the sending lifeline to the receiving lifeline). The label for a synchronous message has the same format as the label for an asynchronous message:

```
<message name> ( <input argument list> )
```

This time, however, the message name must match the name of an *operation* owned by the receiving lifeline.

The input argument list remains optional. If you choose to display the arguments, each one has the same format as before:

```
<input parameter name> = <value specification>
```

Everything I state about the arguments of asynchronous messages is also true for the arguments of synchronous messages. It's worth repeating, however, that the value specification part of this string can be either a literal value or the name of a property that holds a value. (And as I mention earlier, when a sequence diagram serves as a graphical test case specification, it's appropriate for the message arguments to be literal values.)

The sequence diagram in Figure 7.6 shows an excerpt of the interaction displayed in Figure 7.1. The interaction fragment shown here contains two synchronous messages that represent two distinct calls to the *fireThrusters* operation, which is owned by the *Propulsion Subsystem* block.

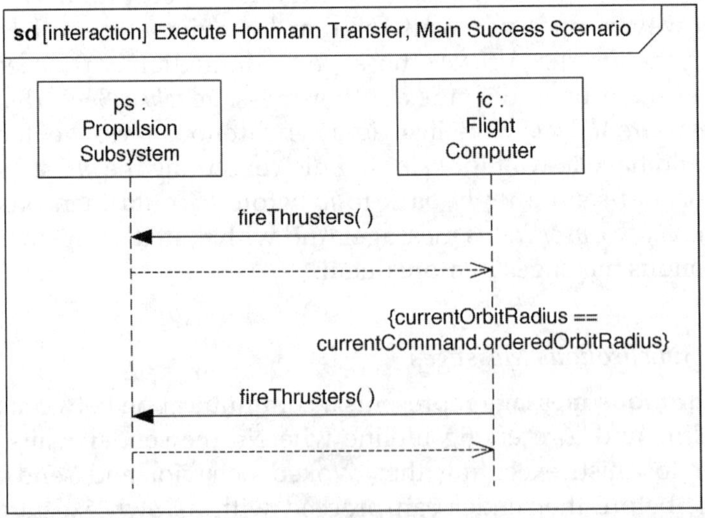

Figure 7.6 *Synchronous messages in an interaction*

These messages have no argument lists displayed. However, all the same key ideas about the arguments of asynchronous messages apply here as well. Any arguments passed in a synchronous message are accessible to the receiving lifeline when it executes the behavior associated with the operation that's being invoked.

The key idea about a synchronous message is that the sender of the message waits for the receiver to finish executing the invoked behavior before the sender proceeds with its own execution. In Figure 7.6, for example, *fc* sends the first *fireThrusters* synchronous message to *ps*. Upon receipt of that message (or at some nondeterministic time thereafter), *ps* begins executing the behavior associated with the *fireThrusters* operation. The *fc* lifeline must wait for that behavior to complete (and it must wait for the reply message that *ps* will send back, marking the completion of that behavior) before *fc* can proceed with its own execution and send the second *fireThrusters* message at some later time.

You might be wondering what that string is that's shown hovering over the *fc* lifeline in between the two *fireThrusters* messages. That string is an example of a state invariant. I talk about these in detail in Section 7.8.3, "State Invariants."

7.5.2.3 *Reply Messages*

A **reply message** represents a communication that marks the end of a synchronously invoked behavior. It's always sent from the lifeline that performed the behavior to the lifeline that invoked the behavior (via a synchronous message earlier in the interaction).

The notation for a reply message is a dashed line with an open arrowhead. Figure 7.6 shows two examples of reply messages—one after each of the corresponding synchronous messages. You can optionally display a label above a reply message. That label has the following format:

```
<assignment target> = <message name>
( <output argument list> ) : <value specification>
```

The message name must match the name of the corresponding synchronous message (which is also the name of the operation that was invoked by that synchronous message).

The value specification after the colon represents the return value of the behavior that just finished executing. Of course, this piece of the string would be present only if the operation that was invoked actually had a declared return type. (Recall from Chapter 3 that operations need not declare a return type.)

The assignment target is an optional piece of information. If shown, it conveys the name of the property that catches the return value—a property owned by the lifeline that invoked the operation (the one receiving the reply message). However, modelers often omit this information.

The output argument list is also optional. If you choose to display the arguments, they appear in a comma-separated list, and each argument has the following format:

```
<output parameter name> : <value specification>
```

The output parameter name is optional; if shown, it must match the name of an output parameter of the operation that was invoked. Modelers often omit this information, too, and simply display the value specification that's being passed back to the caller as an output of the operation that was invoked.

Note

SysML also allows you to optionally display an assignment target for each output argument in the list. But doing so would make the reply message label even more unwieldy than it already is.

Note that showing the reply message itself is optional. The interaction fragment shown in Figure 7.7 is equivalent to the interaction fragment shown in Figure 7.6. Modelers often omit reply messages to con-

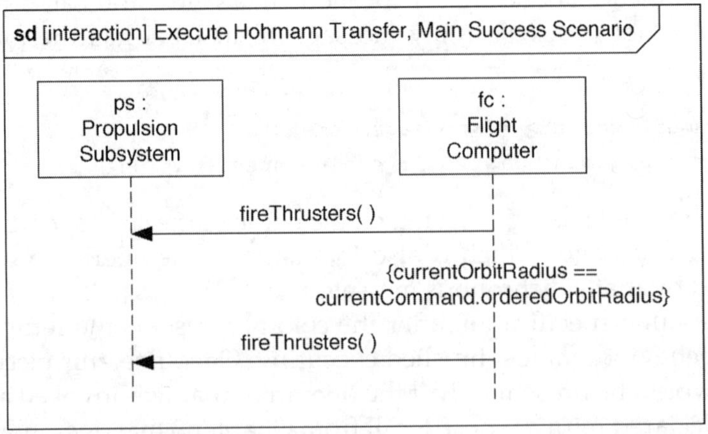

Figure 7.7 *Synchronous messages with implicit reply messages*

serve real estate on a sequence diagram. To be clear, though, a reply message still gets sent to the caller; it's implicit when not shown. The lifeline that sends a synchronous message always receives a reply message upon completion of the invoked behavior, even if that reply message is not shown on the sequence diagram.

7.5.2.4 *Create Messages*

A **create message** represents a communication that creates a new instance within a system—an instance that then participates in the interaction. The notation for a create message is a dashed line with an open arrowhead. The tail end of the message is connected to the sending lifeline (as usual). The arrowhead end of the message is connected to the head of the lifeline that's getting created.

As mentioned, there are six kinds of event occurrences that can appear on lifelines. Thus far, I've discussed two of them: message send occurrences and message receive occurrences. Now it's time for the third: lifeline creation occurrences. A **lifeline creation occurrence** exists on a lifeline at the point where the arrowhead end of a create message meets the head of that lifeline. (A message receive occurrence also exists at that point; it is therefore coincident with the lifeline creation occurrence.)

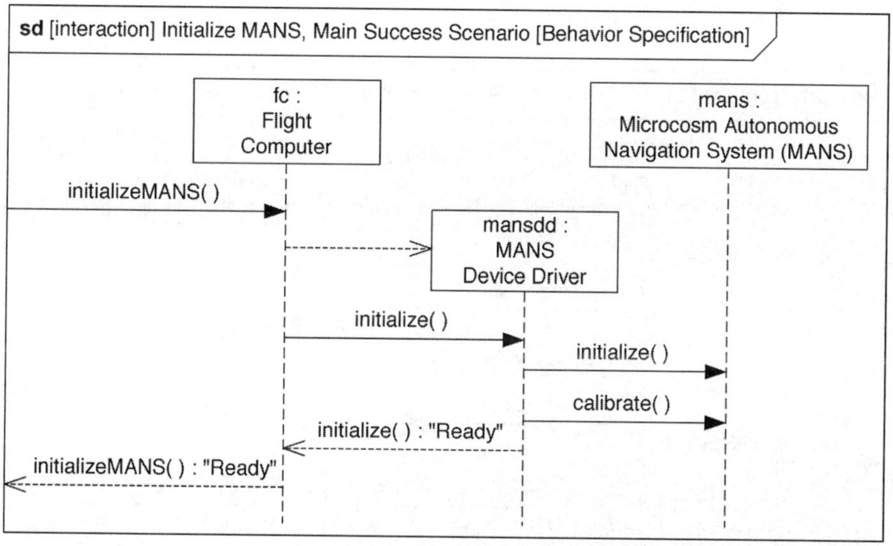

Figure 7.8 *A create message in an interaction*

The sequence diagram in Figure 7.8 displays the *Initialize MANS, Main Success Scenario* interaction. This interaction contains a create message drawn from the *fc* lifeline to the head of a new lifeline, *mansdd*. This conveys that a new instance of the *MANS Device Driver* block gets created at that point in the interaction. The *mansdd* lifeline then receives and sends messages (i.e., participates in the interaction) following its lifeline creation occurrence.

7.6 Destruction Occurrences

The fourth kind of event occurrence that can appear on a lifeline is a lifeline destruction occurrence (or destruction occurrence, for short). A **destruction occurrence** represents the termination of a lifeline and the destruction of the instance within a system that the lifeline represents.

SysML does not define what "destruction" means in any particular context. For a software object, destruction typically refers to the act of freeing the memory that was allocated for that object. For a hardware object, destruction may refer to the removal of an object from the system or to the actual physical destruction of an object (such as pyrotechnic devices that separate a spacecraft from a launch vehicle).

The notation for a destruction occurrence is a cross in the form of an X at the bottom of the lifeline that's getting destroyed (see Figure 7.9). The X may appear at the bottom of a lifeline without any message at-

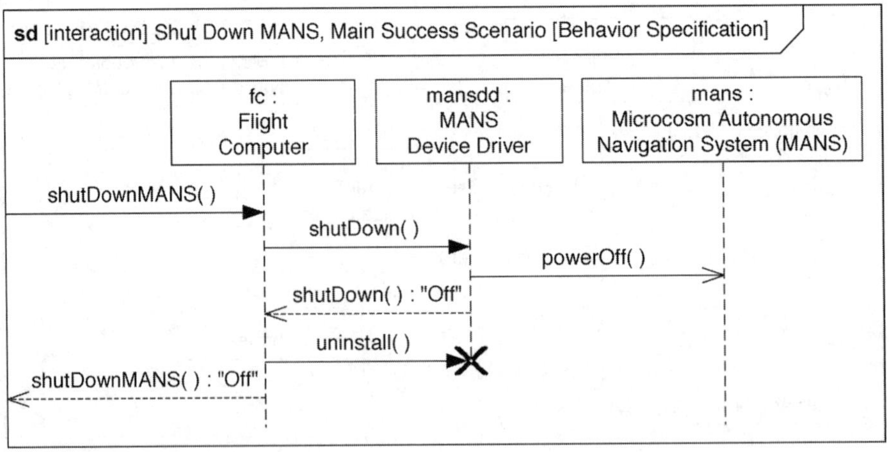

Figure 7.9 *A destruction occurrence in an interaction*

tached to it. This indicates that the lifeline self-terminates at a particular point in the interaction.

The X may also be connected to the arrowhead end of a message. This conveys that the destruction occurrence is the result of the lifeline receiving a special type of message referred to as a **delete message**. The SysML (and UML) specification is oddly silent about the required line style and arrowhead style of a delete message. The language states only that a delete message must end in a destruction occurrence.

The sequence diagram in Figure 7.9 displays the *Shut Down MANS, Main Success Scenario* interaction. This interaction contains a destruction occurrence at the bottom of the *mansdd* lifeline. This particular destruction occurrence is the result of the *mansdd* lifeline receiving an *uninstall* message from the *fc* lifeline. This destruction occurrence conveys that an instance of the *MANS Device Driver* block gets destroyed at that point in the interaction. A destruction occurrence is the last event occurrence that can appear on a lifeline. The *mansdd* lifeline, therefore, can neither send nor receive any messages at any point in the interaction after its destruction occurrence.

7.7 Execution Specifications

The last two kinds of event occurrences that can appear on a lifeline are behavior execution start occurrences and behavior execution termination occurrences. A **behavior execution start occurrence** is generally implicit at the point where a lifeline receives a synchronous or asynchronous message. A **behavior execution termination occurrence** is generally implicit at the point where a lifeline sends a reply message.

It's often useful, however, to eliminate any ambiguity by explicitly conveying to your readers where behaviors begin and end on a lifeline. SysML provides an optional mechanism to do just that: **execution specifications**.

The notation for an execution specification is a thin, vertical rectangle—either white or shaded—that covers a lifeline for the period of time within an interaction when that lifeline is actively executing a behavior. The top of the rectangle explicitly marks a behavior execution start occurrence. The bottom of the rectangle explicitly marks a behavior execution termination occurrence.

Although it's not required by the language, a behavior execution start occurrence is usually coincident with a message receive occurrence.

Therefore, you normally see the arrowhead end of a synchronous or asynchronous message connected to the top of an execution specification. Similarly, a behavior execution termination occurrence is usually coincident with the sending of a reply message, so you normally see the tail end of a reply message connected to the bottom of an execution specification. (But, of course, this applies only for a behavior that was invoked via a synchronous message and not an asynchronous message.)

Figure 7.10 displays an alternative view of the same interaction shown earlier in Figure 7.8. Here, execution specifications are displayed to explicitly convey when the three lifelines begin and end execution of behaviors. The *fc* lifeline executes the *initializeMANS* behavior during the period of time from the receipt of the *initializeMANS* synchronous message to the sending of the corresponding reply message.

While the *initializeMANS* behavior is executing, the *fc* lifeline creates the *mansdd* lifeline and then sends an *initialize* synchronous message, and that causes the corresponding behavior to begin executing on the *mansdd* lifeline. This behavior finishes executing at the point when the *mansdd* lifeline sends a reply message back to the *fc* lifeline.

While the *initialize* behavior is executing on the *mansdd* lifeline, that lifeline sends two synchronous messages—*initialize* and *calibrate*—to

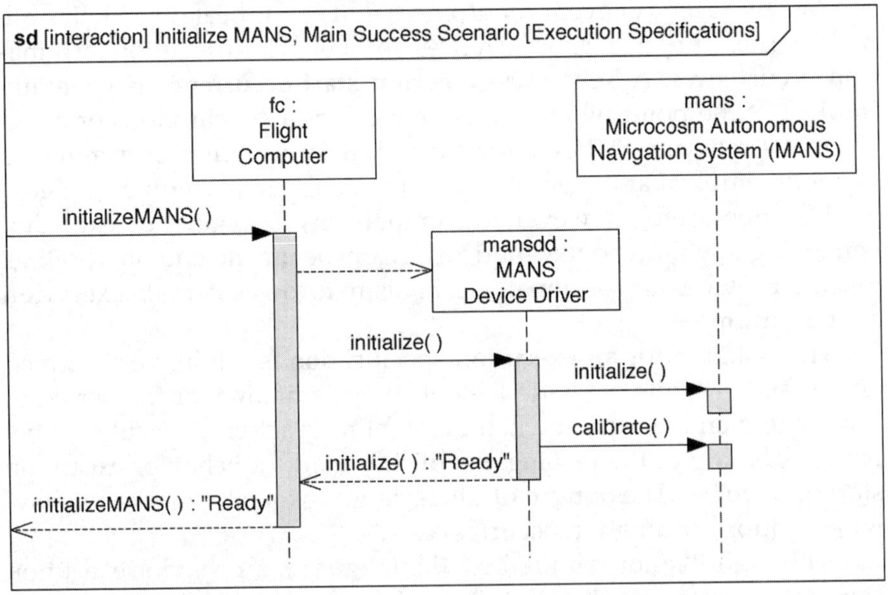

Figure 7.10 *An interaction with execution specifications displayed*

the *mans* lifeline. The receipt of those messages on the *mans* lifeline causes the corresponding behaviors to begin executing.

Note that no reply messages are shown at the points when those behaviors finish executing. Recall from Section 7.5.2.3, "Reply Messages," that it's optional to display a reply message; when not shown, a reply message is implicit at the end of each behavior that was invoked via a synchronous message.

Sometimes a lifeline executes a nested behavior within the context of an outer behavior. You can convey this by using a smaller execution specification that partially overlaps a larger execution specification on the same lifeline. Examples are shown in the interaction displayed earlier in Figure 7.1.

7.8 Constraints

Sequence diagrams allow you to specify various kinds of constraints. In Chapter 3, you learned that a constraint is a Boolean expression that's generally displayed between a pair of curly brackets and can appear on various kinds of SysML diagrams. In the context of interactions, there are three kinds of constraints you will use in daily practice: time constraints, duration constraints, and state invariants.

7.8.1 Time Constraints

Figure 7.11 displays an excerpt of the larger interaction shown in Figure 7.1. This interaction fragment contains at least one example of each of the three kinds of constraints.

The constraint *{currentCommand.executionTime}* in Figure 7.11 is an example of a time constraint. A **time constraint** specifies a required time interval for a *single* event occurrence. That time interval may be a single time value (that is, *min.* and *max.* are equal) or even a property that holds a time value. And the event occurrence that it's attached to can be any one of the six kinds listed at the end of Section 7.4.

The key idea here: When the interaction executes during system operation, it's considered to be a valid execution only if that event occurrence happens within the time interval specified by the time constraint.

The time constraint in Figure 7.11 is attached to the first *fireThrusters* message send occurrence on the *fc* lifeline. This conveys that an execution of this interaction is considered a valid execution only if *fc* sends

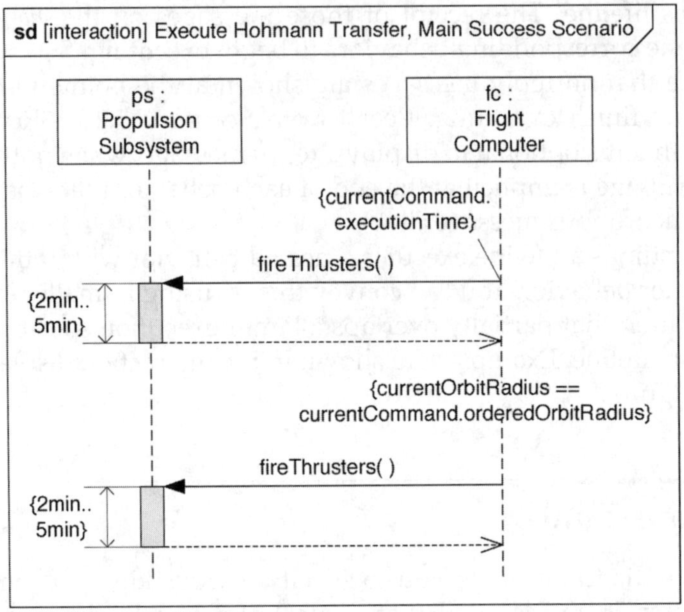

Figure 7.11 *Specifying constraints in an interaction*

the first *fireThrusters* message at the time held in the *executionTime* property (which is a part of the *currentCommand* property, as conveyed by the use of dot notation).

7.8.2 Duration Constraints

The two constraints *{2min..5min}* in Figure 7.11 are examples of duration constraints. A **duration constraint** specifies a required time interval for a *pair* of event occurrences. Again, that time interval may be a single time value or a property that holds a time value. And the pair of event occurrences that it's attached to can be any two of the six kinds listed at the end of Section 7.4.

The key idea here: When the interaction executes during system operation, it's considered a valid execution only if the lapse between that pair of event occurrences falls within the time interval specified by the duration constraint.

Each duration constraint in Figure 7.11 is attached to both the behavior execution start occurrence and the behavior execution termination occurrence for the *fireThrusters* behavior, which is executed by the *ps* lifeline. This conveys that an execution of this interaction is considered a valid execution only if *ps* executes the *fireThrusters* behavior for a minimum duration of two minutes and a maximum duration of 5 minutes (each time that behavior is invoked).

It's also common practice to apply a duration constraint to a message send occurrence and its corresponding message receive occurrence. This constrains the allowable transmission time of the message. When applying a duration constraint for this purpose, you display it in curly brackets floating above or below the constrained message.

7.8.3 State Invariants

The following constraint in Figure 7.11 is an example of a state invariant:

```
{currentOrbitRadius == currentCommand.orderedOrbitRadius}
```

A **state invariant** is a condition that you apply to a specific lifeline at a point preceding (immediately above) a particular event occurrence. That condition must hold true for that lifeline at the moment of that event occurrence in a valid execution of the interaction.

The state invariant in Figure 7.11 is applied to the *fc* lifeline preceding the second *fireThrusters* message send occurrence. This conveys that an execution of this interaction is considered a valid execution only if the value held in the *currentOrbitRadius* property is equal to the value held in the *orderedOrbitRadius* property at the moment when *fc* sends the second *fireThrusters* message.

You can also express a state invariant using the state notation (the round-cornered rectangle) that commonly appears on state machine diagrams. As before, this notation appears on a specific lifeline at a point preceding a particular event occurrence. In this form, however, the state invariant does not contain a Boolean expression; instead, it contains the name of a state that the lifeline must be in at the moment of that event occurrence. However, this notation is meaningful only if the block (whose name appears after the colon in the head of the lifeline) actually owns a state machine behavior having a defined set of states.

The interaction fragment shown in Figure 7.12 is a variant of the one shown in Figure 7.11. In this variant, the *Flight Computer* block owns a state machine behavior (defined somewhere in the model hierarchy), and this behavior contains a state named *At Transfer Orbit Apogee*. This state is named in the state invariant on the *fc* lifeline to convey that the lifeline must be in that state at the moment when it sends the second *fireThrusters* message. If it's not, then the execution of the interaction is invalid.

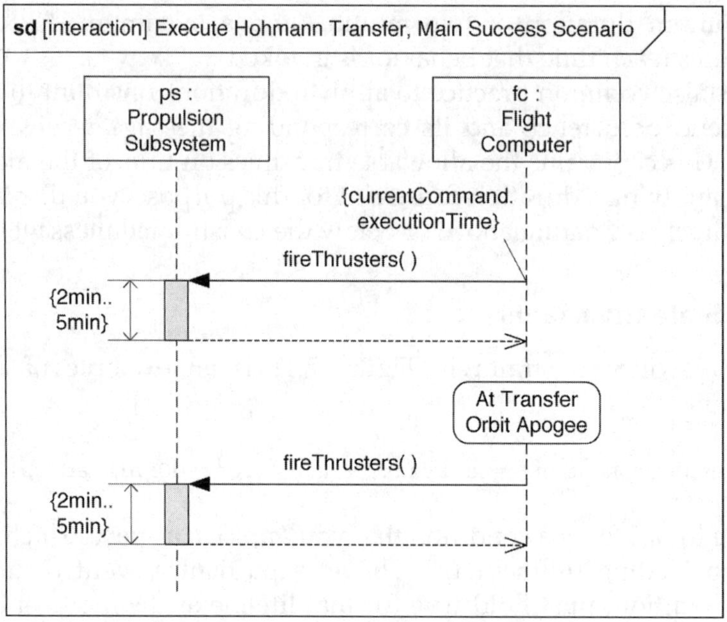

Figure 7.12 *Specifying a state invariant using the state notation*

7.9 Combined Fragments

A **combined fragment** is a mechanism that allows you to add control logic (such as decisions, loops, parallel behaviors) to an interaction. The notation for a combined fragment is a rectangle that appears somewhere within the frame of the sequence diagram. The rectangle is placed over one or more lifelines and encapsulates one or more messages that pass between those lifelines—messages that are subject to the control logic defined by that combined fragment.

You specify the kind of control logic by using a string that appears in a pentagon in the upper-left corner of the rectangle. We call that string an **interaction operator**. SysML defines 11 interaction operators. There are four, however, that you're likely to use in your daily modeling work: *opt*, *alt*, *loop*, and *par*. I discuss each of these in detail in the coming sections.

Before we dive into those details, you should know that each combined fragment is made up of one or more interaction operands (or operands, for short). **Operands** appear as regions within a combined fragment—regions that are separated by dashed lines running horizontally across the rectangle. Each operand (each region) within a combined fragment contains one or more messages that may or may not occur based on the control logic defined by that combined fragment. This will become clearer when you see concrete examples of combined fragments in the sections that follow.

I discuss each of the four common kinds of interaction operators separately in the next four sections, respectively. It's important to know up front that you can nest combined fragments within other combined fragments to create arbitrarily complex control logic. You can see examples of this in the sequence diagram in Figure 7.1 presented earlier.

7.9.1 *Opt* Operator

A combined fragment with an **opt** interaction operator (in the upper-left corner) represents an optional set of event occurrences that could happen during an execution of the interaction if a condition—called the **guard**—evaluates to true. An *opt* combined fragment has exactly one operand (region), so you don't see a dashed line running horizontally across the rectangle. The event occurrences in that one operand either happen, or they don't (based on the evaluation of the guard during system operation).

The guard is a Boolean expression that you display between square brackets near the top of the *opt* combined fragment. You must place the guard on a lifeline directly above the first event occurrence in the combined fragment. Any properties that appear in the Boolean expression must be properties of that lifeline or properties of the block that owns the interaction as a whole.

Figure 7.13 displays an excerpt of the larger sequence diagram shown in Figure 7.1, here focusing on the *opt* combined fragment. This combined fragment contains many event occurrences—specifically, message send occurrences, message receive occurrences, behavior execution start occurrences, and behavior execution termination occurrences. If the guard, *isCommandValid == True*, evaluates to true during an execution of this interaction, then this enclosed set of event occurrences becomes part of that execution. If the guard evaluates to false, then the enclosed set of event occurrences is skipped entirely.

The guard contains the property *isCommandValid*. This is either a property of the *fc* lifeline (which corresponds to a property of the *Flight Computer* block) or a property of the block that owns the interaction as a whole (which is never named on a sequence diagram). You can, however, convey this additional information on a BDD.

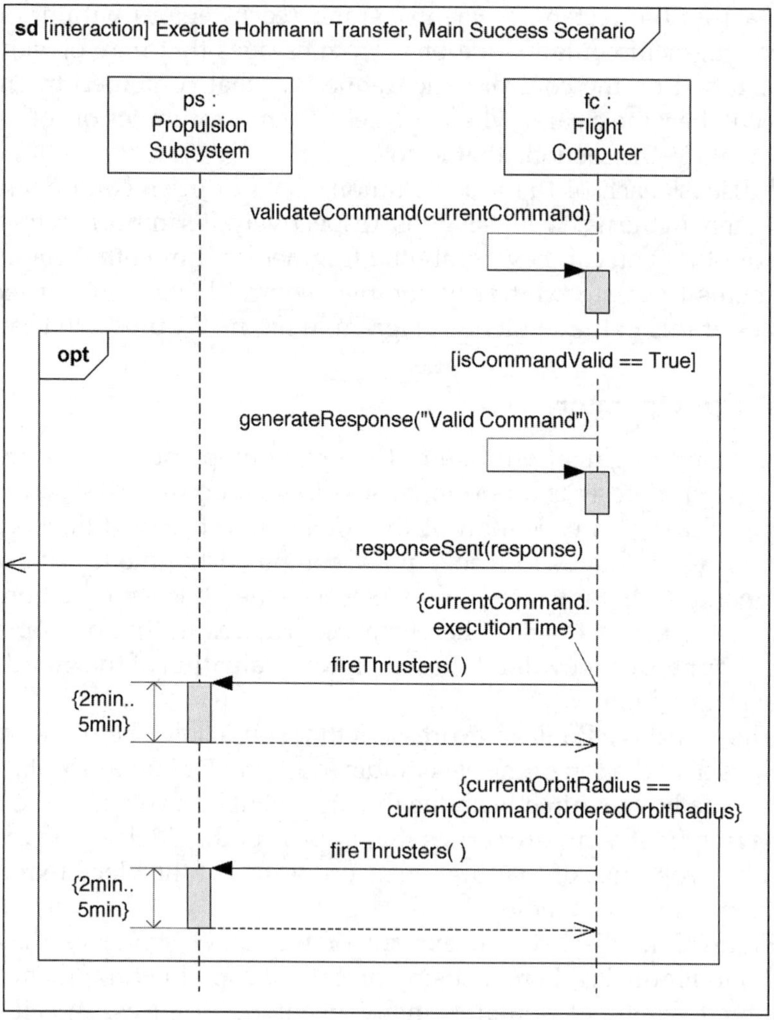

Figure 7.13 *An* opt *combined fragment in an interaction*

7.9.2 *Alt* Operator

A combined fragment with an **alt** interaction operator represents two or more alternative sets of event occurrences that could happen during an execution of an interaction. An *alt* combined fragment must have two or more operands (regions) that contain those alternative sets of event occurrences. As mentioned previously, each operand is separated by a dashed line running horizontally across the rectangle.

Each operand in an *alt* combined fragment has its own guard. At most one of the guards is allowed to evaluate to true, and the set of event occurrences in *that* guard's operand becomes part of the execution; the sets of event occurrences in all of the other operands are skipped entirely. The onus is on you to ensure that the set of guards in an *alt* combined fragment is mutually exclusive. If multiple guards could evaluate to true simultaneously, your model is considered ill formed (a diplomatic way of saying you didn't follow SysML's rules).

You're allowed to use the predefined guard *else* for (at most) one operand. This guard evaluates to true only when all the others evaluate to false. But you aren't required to use the *else* guard. It's possible (and permissible) for none of the guards to evaluate to true (in which case the entire *alt* combined fragment is skipped).

Figure 7.14 shows the *Initialize MANS* interaction, which consists of both the main success scenario (first shown in Figure 7.8) and an error

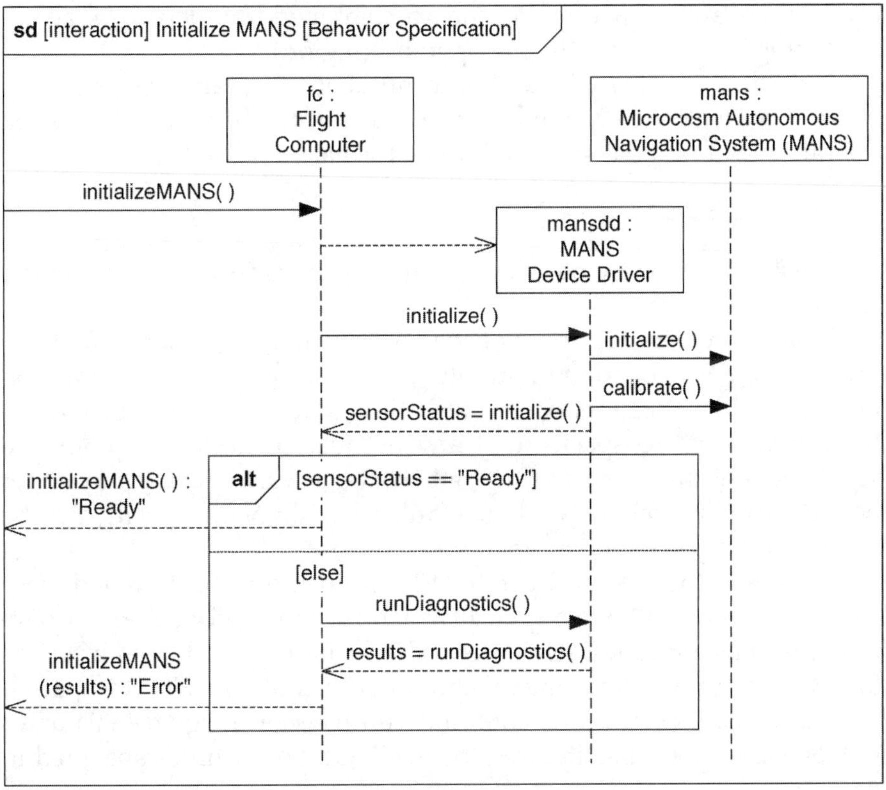

Figure 7.14 *An* alt *combined fragment in an interaction*

scenario. This interaction contains an *alt* combined fragment to model the differences in the endings of these two scenarios.

When the *mansdd* lifeline sends the *initialize* reply message to the *fc* lifeline, the return value of that message is stored in the property *sensorStatus* (which is either a property of the *fc* lifeline or a property of the block that owns the interaction). This property is then read to evaluate the guard in the first operand of the *alt* combined fragment. If that guard evaluates to true, then the *fc* lifeline sends the *initializeMANS* reply message with a return value of *"Ready"* (and all the event occurrences in the other operand are skipped). If that guard evaluates to false, then the *else* guard instead evaluates to true, and the event occurrences in the second operand become part of the execution.

7.9.3 *Loop* Operator

A combined fragment with a **loop** interaction operator represents a set of event occurrences that could happen multiple times during a single execution of an interaction. Like an *opt* combined fragment, a *loop* combined fragment has exactly one operand (region).

You can specify a *min.* and *max.* number of iterations of the loop between parentheses immediately to the right of the *loop* interaction operator. That range is specified in the format

```
(<min.>, <max.>)
```

If *min.* and *max.* are equal, you can simply display a single number as a shorthand form.

Note that this range does not specify how many iterations *will* occur during a single execution of an interaction. Rather, it's a constraint on how many iterations *can* occur (and still result in a valid execution of the interaction). To specify that any number of iterations would be valid, you set the range to (0, *), where the asterisk means "no upper bound." And, in fact, (0, *) is the default range if none is specified to the right of the *loop* interaction operator.

As usual, the operand contained within the *loop* combined fragment is allowed to have a guard (which is displayed between square brackets near the top of its operand). The guard is evaluated after the loop has iterated at least the *min.* number of times specified in parentheses. Once it has, the loop continues, until either the guard evaluates to false or the loop has iterated the *max.* number of times specified in parentheses.

Figure 7.15 displays an excerpt of the larger sequence diagram shown in Figure 7.1, with one difference: The *loop* combined fragment

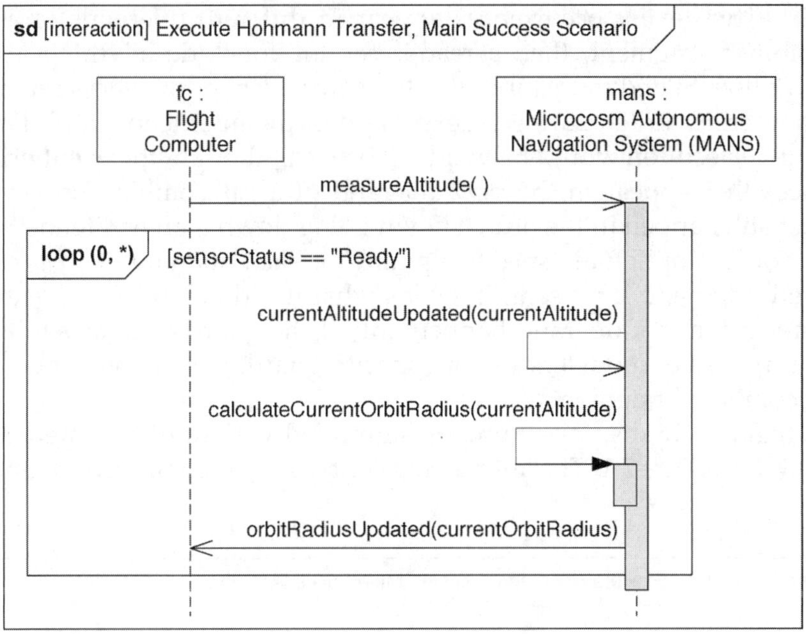

Figure 7.15 *A* loop *combined fragment in an interaction*

shown here has a guard specified. This combined fragment encapsulates the message send and message receive occurrences for three messages, as well as a behavior execution start and a behavior execution termination occurrence. This entire sequence of occurrences could happen any number of times during a single execution of the *Execute Hohmann Transfer, Main Success Scenario* interaction (as conveyed by the range displayed after the *loop* interaction operator). This loop will terminate, however, the moment the guard *sensorStatus == "Ready"* evaluates to false.

7.9.4 *Par* Operator

A combined fragment with a **par** interaction operator represents two or more sets of event occurrences that happen in parallel with each other during an execution of an interaction. Like an *alt* combined fragment, a *par* combined fragment has two or more operands (regions), which contain those concurrent sets of event occurrences.

In Section 7.4, you learned that an event occurrence that appears higher on a lifeline happens before one that appears lower on the same lifeline. However, the presence of a *par* combined fragment changes that. If two event occurrences appear in different operands of a *par* combined fragment, then a reader cannot conclude anything about their order. Stated more formally, those two event occurrences can happen in either order during an execution of the interaction, and the resulting execution would be valid. To be clear, though, the event occurrences that appear in the *same* operand of a *par* combined fragment must still happen in the order shown going down a given lifeline.

You can optionally specify a guard for each operand of a *par* combined fragment. If present, it means what it did before: The event occurrences in that operand happen only if the guard evaluates to true. With that said, modelers seldom specify guards for the operands of a *par* combined fragment.

Figure 7.16 displays a greatly simplified variant of the interaction shown in Figure 7.1. This interaction contains a *par* combined fragment.

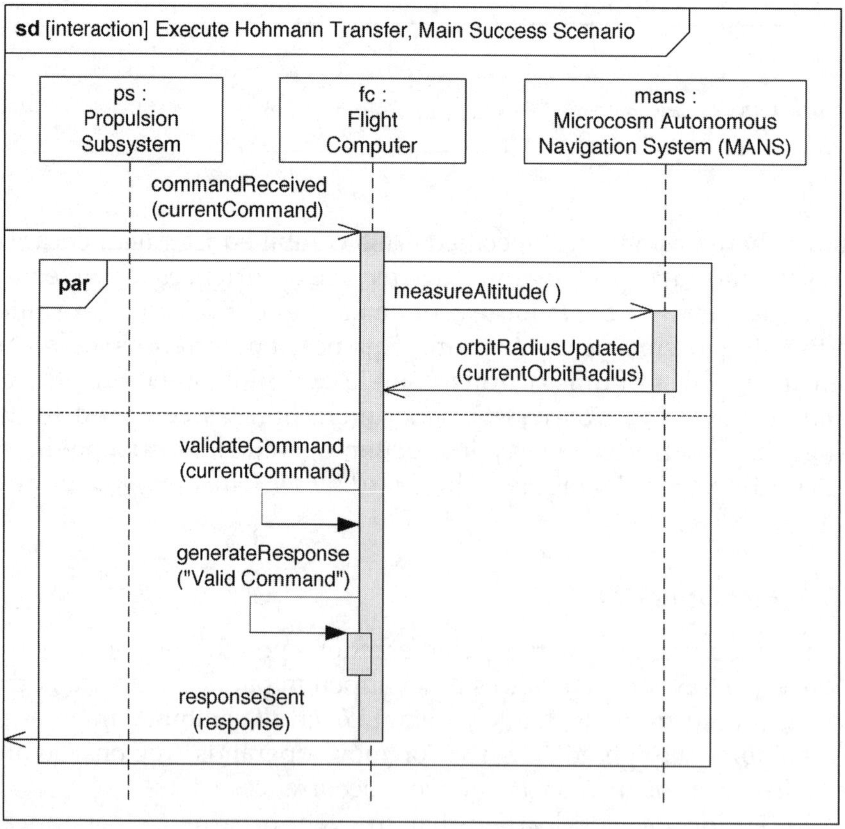

Figure 7.16 *A* par *combined fragment in an interaction*

In the first operand, there are two event occurrences on the *fc* lifeline. In the second operand, there are eight event occurrences on the *fc* lifeline. The *par* combined fragment conveys that these sets of event occurrences happen in parallel with each other.

It's possible that the two event occurrences in the first operand could happen before the eight event occurrences in the second operand. It's also possible that all eight event occurrences in the second operand could happen before the two event occurrences in the first operand. And, of course, it's possible that these two sets of event occurrences could be interleaved in some nondeterministic way. That's the nature of concurrent behaviors; their relative order cannot be determined ahead of time.

7.10 Interaction Uses

In Chapter 6, you learned that you can decompose a high-level activity into lower-level behaviors—behaviors that get invoked via call behavior actions. Similarly, you can decompose a high-level interaction into lower-level behaviors—behaviors that get invoked via an element called an **interaction use**.

The notation for an interaction use is a rectangle that appears somewhere within the frame of the sequence diagram. A pentagon appears in the upper-left corner of the rectangle containing the string *ref*, which conveys that the interaction use is a reference to another interaction that you've defined somewhere in your model hierarchy. The name of that referenced interaction appears inside the rectangle. The rectangle must be placed over the lifelines that participate in that referenced interaction. (The participating lifelines disappear behind the rectangle.)

Modelers generally add an interaction use to an interaction for one of two reasons:

- To factor out a subset of event occurrences that are common to several high-level interactions and put that subset in one place in a single, lower-level interaction
- To decompose a complex set of event occurrences (in a high-level interaction) into a more readable sequence of lower-level interactions

Figure 7.17 displays the *Initialize MANS, Main Success Scenario* interaction originally shown in Figures 7.8 and 7.10. In this sequence dia-

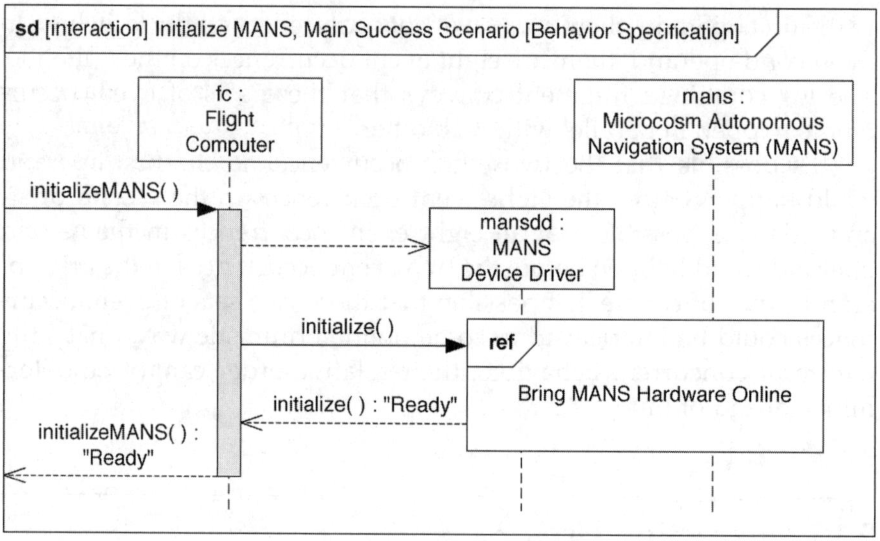

Figure 7.17 *An interaction use with actual gates*

gram, however, I've factored out a subset of event occurrences, placed them in a lower-level interaction named *Bring MANS Hardware Online*, and added an interaction use to show the lower-level interaction getting invoked.

Figure 7.18 displays the *Bring MANS Hardware Online* interaction, which contains the subset of event occurrences that I factored out of the interaction in Figure 7.17. Referenced interactions can themselves con-

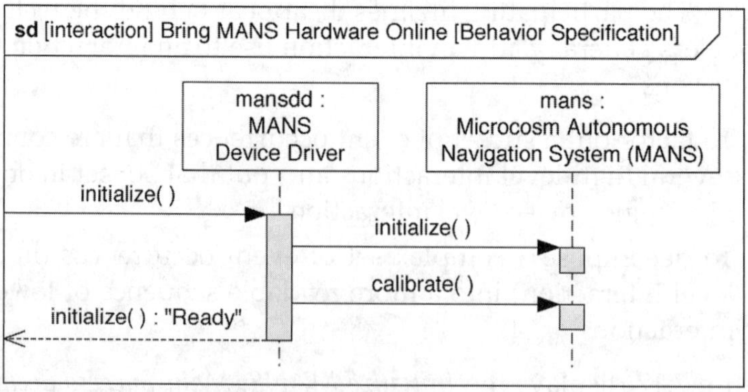

Figure 7.18 *An interaction with formal gates*

tain interaction uses, letting you create an arbitrarily deep decomposition of behaviors.

If an interaction use has messages entering it or leaving it (as shown in Figure 7.17), then the interaction it references must have matching messages coming in from the frame or going out to the frame (as shown in Figure 7.18). Formally, SysML states that a message enters (or leaves) an interaction use at an **actual gate**. And SysML states that a message enters (or leaves) the referenced interaction at a **formal gate**.

There's no distinct notation for actual gates or formal gates; they are implicit at the points where messages intersect an interaction use or the diagram frame, respectively. The key idea: There must be a one-to-one correspondence between the actual gates and formal gates. (As long as you remember to make the messages match, though, you can get by without knowing this jargon.)

Summary

A sequence diagram conveys information about a system's behavior over time, with a focus on the communications that occur among specific parts within a system. Modelers often create a sequence diagram to model a test case—a single path of execution through a use case with specified input values and expected output values. An important strength of a sequence diagram is its ability to completely and unambiguously specify a system behavior. It conveys all three essential pieces of information: the order of the behaviors that occur, which structure performs each behavior, and which structure invokes each behavior. For this reason, sequence diagrams often serve as inputs into the development stage of the system life cycle.

Chapter 8

State Machine Diagrams

State machine diagrams are the last of the three kinds of SysML diagrams that you can use to express information about a system's dynamic behavior. You can display various kinds of states on a state machine diagram, and you can specify four kinds of events to trigger transitions among those states in a running system. SysML also lets you use orthogonal regions to model concurrent state-based behavior.

8.1 Purpose

A **state machine diagram** is a kind of behavior diagram; like an activity diagram and a sequence diagram, it presents a dynamic view of a system. Unlike an activity diagram and a sequence diagram, a state machine diagram focuses attention on how a structure within a system changes state in response to event occurrences over time.

The behavior displayed on a state machine diagram most often represents a block's **classifier behavior**, a formal term that refers to the behavior that begins executing the moment a block is instantiated and generally finishes executing when that instance is destroyed. It's also legal for a state machine behavior to be associated with a single operation or reception of a block, but this is rare.

A state machine diagram is well suited to serve as a detailed design artifact (that is, an input into development). Like a sequence diagram, a state machine diagram is a precise and unambiguous specification of behavior. Many commercial-grade modeling tools let you autogenerate production-quality source code based on the behavior displayed on a state machine diagram.

The drawback to a state machine diagram is that its use is limited to describing the behavior of blocks that actually have defined states (that is, they exhibit state-based behavior in response to event occurrences). Not all blocks have such defined states.

8.2 When Should You Create a State Machine Diagram?

Because a state machine behavior most often serves as a block's classifier behavior, you can create a state machine diagram to describe the behavior of a block at *any* level in the system hierarchy (such as the system of interest itself, a subsystem, or a single component). Therefore, you can potentially create a state machine diagram at any point in the system life cycle.

8.3 The State Machine Diagram Frame

The diagram kind abbreviation for a state machine diagram is *stm*. The only allowable model element type for a state machine diagram is *stateMachine*. The frame of a state machine diagram always represents a single state machine that you've defined somewhere in your system model.

A **state machine** is itself a model element; like an interaction and an activity, it's a kind of behavior. Like an interaction, an activity, a block, and a package, a state machine is also a kind of namespace. It can therefore contain a set of named elements—specifically, vertices and transitions—within the model hierarchy. Those contained elements can appear within the frame of the associated state machine diagram.

A key idea here is that *state machine* and *state machine diagram* are not synonyms. When I use the term *state machine* throughout this chapter, I'm referring to a model element and not the diagram it's displayed on.

The diagram header in Figure 8.1 tells us that the frame of this state machine diagram represents the state machine named *Attitude Control*, which is defined somewhere in the system model. The elements that appear within the frame—the vertices and transitions—are contained within (nested under) this state machine within the model hierarchy.

The name of this diagram is "Attitude Control Subsystem Classifier Behavior." This name conveys the purpose of the diagram: It displays the classifier behavior of the *Attitude Control Subsystem* block. To be clear, you're not required to specify on the state machine diagram the name of the block that the state machine is associated with. However, if you feel it adds value, the diagram name in the header is the only place on the diagram where you can provide that information.

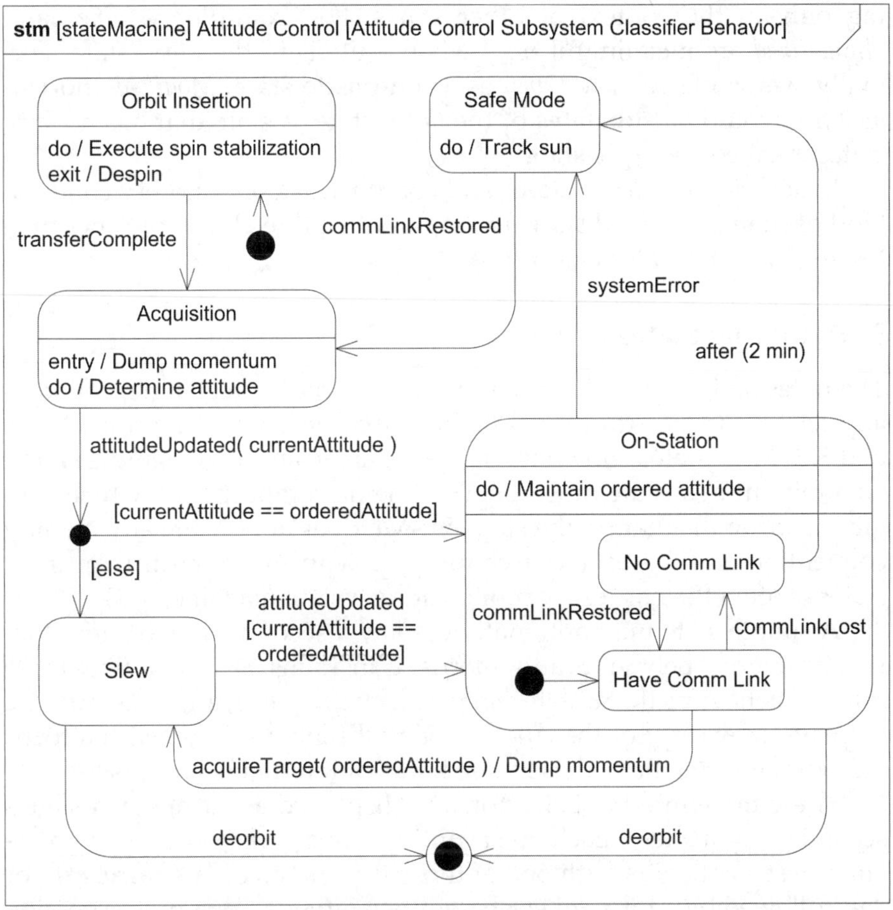

Figure 8.1 *A sample state machine diagram*

8.4 States

A system (or a part within a system) sometimes has a defined set of states in which it can exist during system operation. The concept of **state** is difficult to define formally but easy to infer from real-world examples. One of the simplest examples is a lamp that turns on and off via a pull chain. The lamp has two defined states: *On* and *Off*. The state that it's in at a given moment determines how it will respond to event occurrences (such as unscrewing the light bulb, pulling the chain, or knocking the lamp over).

Software objects, too, can have a defined set of states. A file, for example, can exist in the following states: *Open*, *Closed*, *Modified*, *Unmodified*, *Encrypted*, *Unencrypted*, and others. Sometimes states have meaning only in the context of other states. For example, *Modified* and *Unmodified* are meaningful only when a file is in the *Open* state. Formally, we would refer to *Open* as a **composite state**; *Modified* and *Unmodified* would be **substates** of the *Open* state. A state that has no substates is called a **simple state**.

In addition to simple states and composite states, another common kind of state is the **final state**. Examples of all three kinds are shown in Figure 8.1. Let's take a closer look.

8.4.1 Simple States

The notation for a simple state is a round-cornered rectangle, colloquially referred to as a **round-angle**. The *Orbit Insertion*, *Acquisition*, *Slew*, and *Safe Mode* states shown in Figure 8.1 are examples of simple states. At a minimum, a state must display a name compartment, which contains a string that names the state. SysML does not dictate any naming conventions for states. (And coming up with meaningful names for states is often the hardest part of creating a state machine.)

A simple state may optionally display a second compartment that lists its internal behaviors and internal transitions. SysML defines three internal behaviors that a state can perform: *entry*, *exit*, and *do*. Figure 8.2 displays an excerpt of the *Attitude Control* state machine, with a focus on two states that have these three kinds of internal behaviors.

These three internal behaviors are displayed as strings in a state's second compartment. Each one is optional in a given state. When present, the string begins with one of those three keywords—*entry*, *exit*, or *do*—followed by a forward slash and then either an opaque expression

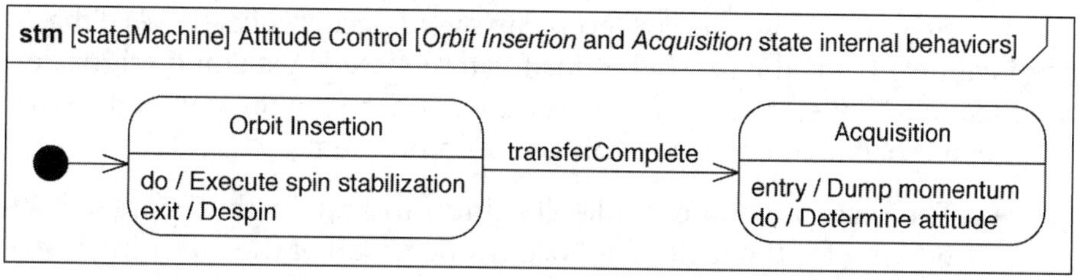

Figure 8.2 Entry, exit, *and* do *behaviors in simple states*

or the name of a behavior that you've created somewhere in the system model.

Opaque expressions are language-specific statements of behavior. I discuss these in greater detail in Section 6.5, "Actions: The Basics," in the context of activity diagrams. If you instead specify the name of a behavior that exists in your model, it can be any one of the three kinds defined in SysML: an activity, an interaction, or another state machine. (It's rare, though, for an *entry, exit*, or *do* behavior to be another state machine.)

A state's *entry* behavior, if present, is the first behavior executed upon entering that state. The *entry* behavior is regarded as **atomic** (uninterruptible). This means that it's guaranteed to finish executing before the state machine can process a new event occurrence (and potentially transition out of that state).

A state's *exit* behavior, if present, is the last behavior executed before leaving that state, which happens when an event occurrence causes the state machine to transition to a new state. Like an *entry* behavior, an *exit* behavior is regarded as atomic. It's guaranteed to finish executing; no new event occurrence can interrupt its execution.

It's important to understand that a state machine may rest in a given state for some nondeterministic period before transitioning to a new state. This means that other behaviors can execute in between a state's *entry* and *exit* behaviors.

A state's *do* behavior, if present, begins executing upon entering the state, immediately following the state's *entry* behavior. The key difference between the *do* behavior and the *entry* behavior is that the *do* behavior is **nonatomic**; its execution *can* be interrupted by a new event occurrence that causes a transition to a new state. A *do* behavior therefore continues executing until either it's interrupted by an event occurrence (and the state machine transitions to a new state) or it terminates on its own (which may happen before the next event occurs).

If a *do* behavior is interrupted by an event occurrence, the *do* behavior is aborted and the *exit* behavior for that state is executed (right before leaving that state). If a *do* behavior terminates on its own before the next event occurs, one of two things could happen:

- The state machine could continue to rest in that state while waiting for the next event occurrence, if all of the outgoing transitions require a trigger.
- The state machine could immediately transition to a new state, if there's an outgoing transition that doesn't require a trigger.

I discuss the concept of a trigger in detail in Section 8.5, "Transitions."

Note

A *do* behavior is formally called a ***do* activity** in SysML. I find this term misleading, though, because a *do* behavior could be an interaction or a state machine, and not necessarily an activity. You need to be aware of this in case you ever need to dive into the SysML specification (or UML specification) to learn more about state machines.

8.4.2 Composite States

The notation for a composite state is the same as the one for a simple state: a round-cornered rectangle. Like a simple state, a composite state has a name compartment and a compartment where you can display the optional *entry*, *exit*, and *do* behaviors. The difference is that a composite state has nested substates, which you would display in a third compartment (below the second compartment).

There are also several similarities between a composite state in a state machine and a state machine as a whole. When a composite state is inactive, all its substates are inactive, too. When a composite state is active, exactly one of its substates is also active.

While active, a composite state can transition from one substate to another in response to event occurrences. Transitions between substates behave in exactly the same manner as transitions between states. (You can read more on this in Section 8.5.)

Figure 8.3 shows an excerpt of the larger *Attitude Control* state machine in Figure 8.1—here, with a focus on the composite state *On-*

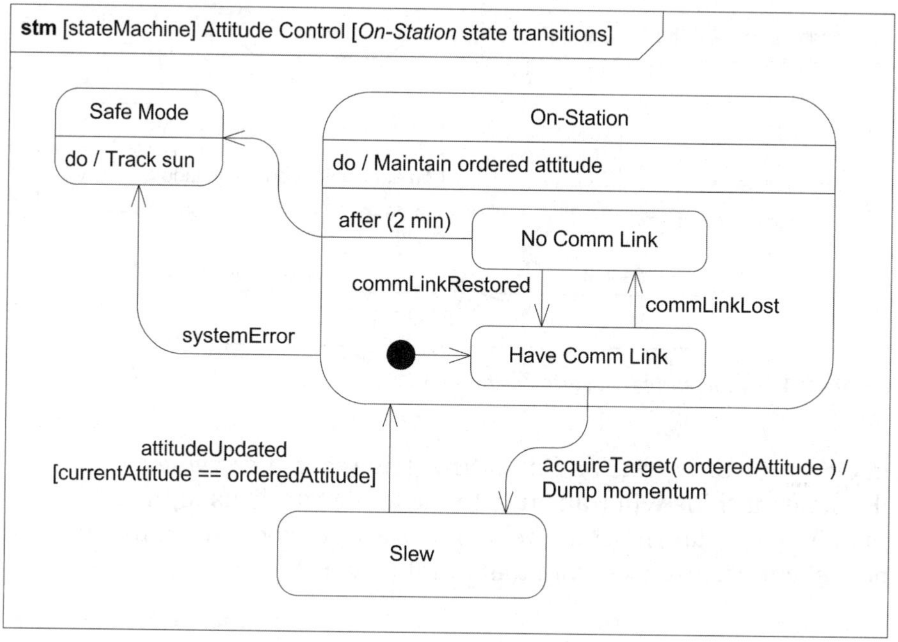

Figure 8.3 *A composite state in a state machine*

Station. This composite state has two substates: *Have Comm Link* and *No Comm Link*. While *On-Station* is active, it can transition between its two substates in response to *commLinkRestored* and *commLinkLost* event occurrences.

Transitions out of a composite state can originate either from the boundary of the composite state or from a specific nested substate. Figure 8.3 shows examples of both cases. Transitions that originate from the boundary of a composite state can fire when an event occurs, no matter which substate the composite state is in at a given moment. Transitions that originate from a specific nested substate can fire—when an event occurs—only if the composite state happens to be in that substate at that moment. (In Section 8.5, I discuss the general rules governing when a transition fires.)

8.4.3 Final States

The notation for a **final state** is a small, filled-in circle surrounded by a larger circle. Figure 8.4 shows an excerpt of the *Attitude Control* state machine shown earlier in Figure 8.1, this time with a focus on the transitions leading to the final state. If the *Attitude Control* state machine is

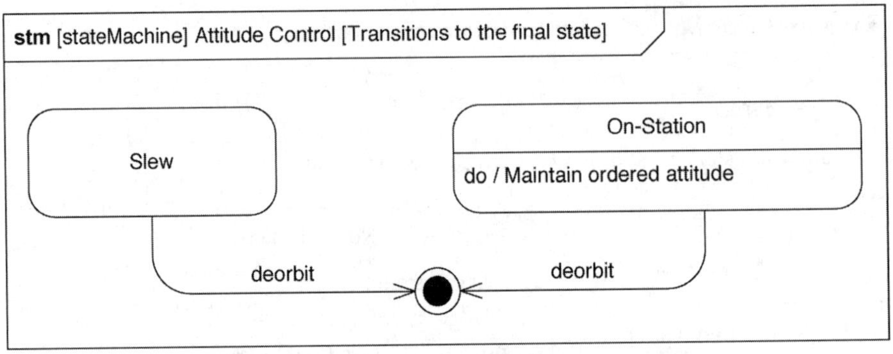

Figure 8.4 *A final state in a state machine*

in either the *Slew* state or *On-Station* state when a *deorbit* event occurs, the state machine will transition to the final state. This represents completion of the state machine behavior as a whole; it will not react to any new event occurrences from that point forward.

8.5 Transitions

A **transition** represents a change from one state to another. What's less intuitive is that it can also represent a change from one state back to itself—what SysML calls a **self-transition**. The notation for a transition is a solid line with an open arrowhead drawn from a source vertex to a target vertex (where source and target may be the same vertex, in the case of a self-transition).

Most often, the source and target vertices are states. However, they can also be pseudostates, which I discuss in detail in Section 8.6, "Pseudostates." For now, we focus strictly on transitions between states.

Each transition can specify three optional pieces of information: a trigger, a guard, and an effect. When present, these pieces of information appear in a single string that floats above or below the transition. The format of that string is as follows:

```
<trigger> [<guard>] / <effect>
```

The trigger must match the name of an event that you've defined in your system model. SysML defines four types of events: signal events, call events, time events, and change events, as discussed in detail in

Section 8.5.2. For now, it's sufficient to know that an instance of an event (during system operation) is called an **event occurrence**, and an event occurrence can **trigger** a transition between states.

The guard is a Boolean expression always expressed between square brackets. That expression evaluates to true or false at any given moment. A transition fires only if its guard is true at the moment when the state machine receives an event occurrence that matches its trigger. If its guard is false at that moment, then the transition does not fire and that event occurrence is consumed without any resulting change in state.

The effect is a behavior that gets executed during the transition. Like an *entry*, *exit*, and *do* behavior, an **effect** can be either an opaque expression or the name of a behavior that you've defined somewhere in the system model. And that behavior can, once again, be an activity, an interaction, or another state machine. (Most often, though, it is an activity or an interaction.)

The transition from the *Offline* state to the *Online* state in Figure 8.5 has a trigger, a guard, and an effect. The trigger is the event named *startUp*, which passes a value into the state machine via the argument *sensorID*. This particular event happens to be a call event (which I discuss shortly), but the discussion of transitions in this section applies equally to the other three types of events.

This transition has the guard *isPowerAvailable == True*. If this guard evaluates to false at the moment when the *startUp* event occurs, then that *startUp* event occurrence is consumed and the state machine remains in the *Offline* state. If this guard evaluates to true at the moment when the *startUp* event occurs, then the transition fires (the state machine transitions from the *Offline* state to the *Online* state).

This transition has an effect named *calibrate*. This behavior can be an activity, an interaction, or another state machine that exists somewhere in the system model. A state machine diagram, unfortunately,

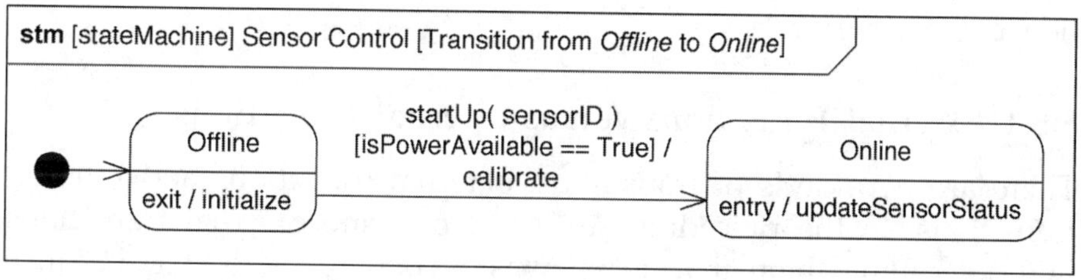

Figure 8.5 *A transition with a trigger, a guard, and an effect*

does not provide a way to specify which kind of behavior it is. The key point is that this effect gets executed when the transition fires (as described in the preceding paragraph).

A transition effect is part of a larger sequence of behaviors that executes when a transition fires. We call that sequence of behaviors the run-to-completion step. The **run-to-completion step** consists of the following behaviors in the order listed:

- The *exit* behavior of the source state
- The *effect* specified for the transition itself
- The *entry* behavior of the target state

This entire sequence of behaviors is regarded as atomic and instantaneous. It's guaranteed to finish executing; no new event occurrence can interrupt any of the behaviors in this sequence. A state machine can receive and process a new event occurrence only after the entire run-to-completion step finishes executing and the state machine is at rest in the new (target) state.

Systems engineers know that no real-world system behavior truly executes in zero time. However, when you choose to model a behavior as part of a run-to-completion step—as an entry behavior, a transition effect, or an exit behavior—you are asserting to your stakeholders that the system will be implemented so that the behavior will be uninterruptible and that no other behavior will run concurrently during its execution, creating the illusion that it runs in zero time.

Keep in mind that each of the three behaviors in the run-to-completion step is optional; any combination of the three can be present for a given transition between two states. When present, however, they get executed in the order shown when that transition fires.

When the state machine shown in Figure 8.5 transitions from the *Offline* state to the *Online* state, the *initialize* behavior gets executed, followed by the *calibrate* behavior, and then the *updateSensorStatus* behavior. When that entire run-to-completion step has finished executing, the state machine will be at rest in the *Online* state and ready to receive a new event occurrence.

8.5.1 External Transitions versus Internal Transitions

There are two kinds of transitions: external transitions and internal transitions. In the preceding section, I describe external transitions without calling them that. (That was a good place to start because they're the more common kind.) To be clear, whenever you see an arrow drawn from one state to another (or from one state back to itself), you're looking at an **external transition**. And everything I state about them in the preceding section applies.

The string format for an internal transition is the same as for an external transition. However, the string for an **internal transition** is displayed in the second compartment of a state (along with the three optional internal behaviors); in contrast to an external transition, the string for an internal transition is not displayed next to an arrow.

That difference in notation reflects the conceptual difference between internal transitions and external transitions: When an internal transition in a state fires, the state machine doesn't leave that state as a result of the transition. And this implies another key difference between internal transitions and external transitions: When an internal transition in a state fires, neither the *exit* behavior nor the *entry* behavior for that state, if present, gets executed. In short, when an internal transition fires, the only behavior that gets executed is the *effect* specified for that internal transition.

Figure 8.6 displays a refinement of the *Sensor Control* state machine from Figure 8.5. In this version, the *Online* state has an internal transition. The trigger for this internal transition is a **change event**:

```
when (availableMemory < dataPerOrbit)
```

This transition has no guard. (If it did, it would be shown in square brackets after the trigger, as it is for external transitions.) Therefore,

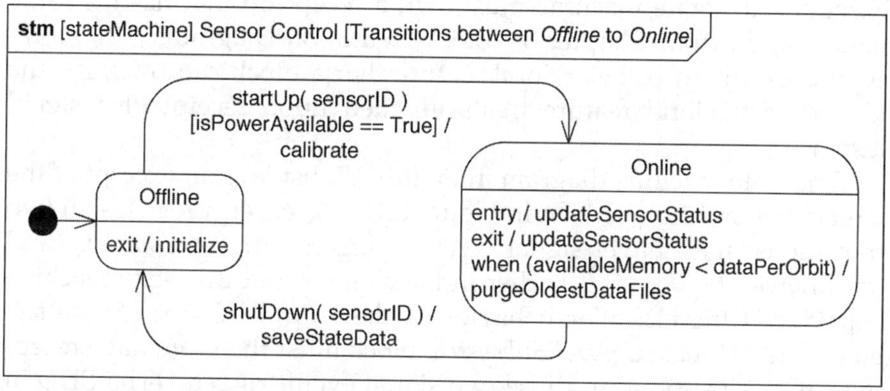

Figure 8.6 *A state with an internal transition*

when the specified change event occurs—that is, when the Boolean expression in parentheses becomes true—this internal transition fires. And when it does, the only behavior that gets executed is the effect specified for the transition: the *purgeOldestDataFiles* behavior.

8.5.2 Event Types

As mentioned earlier in the chapter, SysML defines four types of events: signal events, call events, time events, and change events. Before I discuss the details of each type, it's important to understand a fundamental point about events generally: An **event** is an element of definition within a system model—an element that defines a type of **occurrence** that can trigger a behavior within an operational system. A defined event could potentially occur multiple times during system operation. And each time an event occurs, the occurrence could potentially trigger a new execution of a behavior.

This is true in the context of activities and interactions, as discussed in those chapters. And it remains true in the context of state machines. The difference here is that the focus shifts to how event occurrences trigger transitions between states—transitions that potentially result in behaviors getting executed.

8.5.2.1 Signal Events

A **signal event** represents the receipt of a signal instance by a target structure that's receptive to it; the target in this context would be the structure that's executing a state machine behavior. If the state machine has a transition with a signal event trigger, then the structure that's executing the state machine must own a **reception** that has the same name. Recall from Chapter 3, "Block Definition Diagrams," that a reception is a kind of behavioral feature that a block can own; it's the kind of behavioral feature that's invoked upon receipt of a signal instance.

The state machine diagram in Figure 8.7 displays an excerpt of the larger state machine shown in Figure 8.1. This excerpt focuses on two transitions that have signal event triggers: *transferComplete* and *commLinkRestored*. I state earlier in the chapter that this state machine behavior is the classifier behavior of the *Attitude Control Subsystem* block. The *Attitude Control Subsystem* block must therefore have receptions that correspond to those two signal event triggers. (The BDD in Figure 8.7 conveys this correspondence explicitly.)

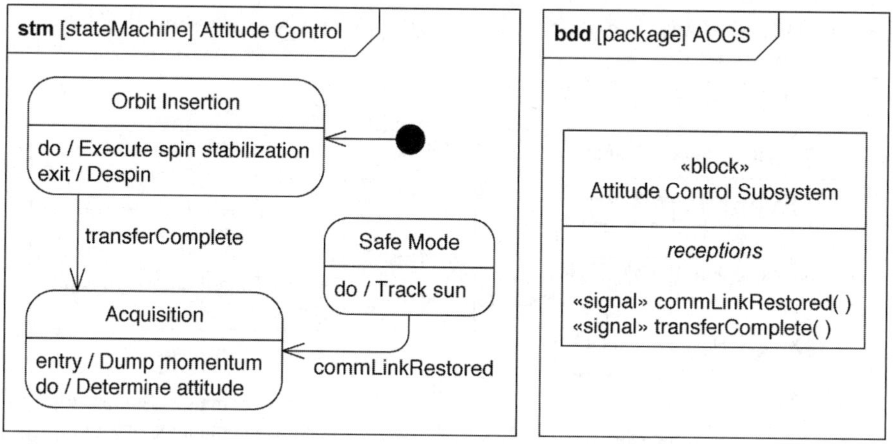

Figure 8.7 *Signal event triggers and corresponding receptions*

If a reception owns parameters (which are shown in parentheses to the right of its name), then the corresponding signal event in the state machine can specify an argument for each of those parameters. The arguments that are passed into a state machine via signal events can then be used both in guards and in behaviors that the state machine invokes. The signal events (and receptions) in Figure 8.7 do not specify any arguments (or parameters). However, the *attitudeUpdated* signal event in Figure 8.1 does specify an argument, *currentAttitude*. That argument is later accessed to evaluate the guard, *currentAttitude == orderedAttitude*.

8.5.2.2 *Call Events*

A **call event** represents the receipt of a request to invoke an operation in a target structure—a request that's sent from a calling structure. The target in this context would be the structure that's executing a state machine behavior. If the state machine has a transition with a call event trigger, then the structure that's executing the state machine must own an operation that has the same name. Recall from Chapter 3 that an **operation** is a kind of behavioral feature that a block can own; it's the kind of behavioral feature that's generally invoked by a synchronous call (one wherein the caller waits for the behavior to finish).

The state machine diagram in Figure 8.8 displays an excerpt of the larger state machine shown in Figure 8.1. This excerpt contains a transition with a call event trigger: *acquireTarget*. The *Attitude Control*

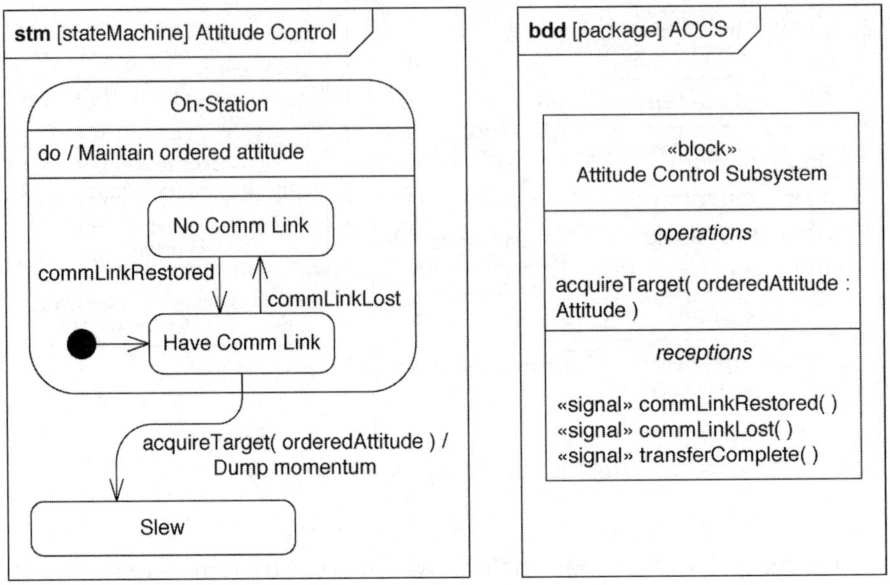

Figure 8.8 *A call event trigger and a corresponding operation*

Subsystem block, which owns this state machine behavior, must have an operation that corresponds to that call event trigger. The BDD in Figure 8.8 conveys this correspondence explicitly.

The *acquireTarget* operation owns the parameter *orderedAttitude* of type *Attitude*. The corresponding call event in the state machine, therefore, specifies an argument for that parameter. In this particular example, the parameter and its corresponding argument have the same name. SysML doesn't require that, but it's a good idea in daily practice.

As with signal events, the arguments that are passed into a state machine via call events can be used both in guards and in behaviors that the state machine invokes. The excerpt shown in Figure 8.8 omits those guards. Figure 8.1, though, shows many guards that access the *orderedAttitude* argument.

You may have noticed the similarity in the notations for call event triggers and signal event triggers. They're identical, in fact. Your readers therefore have no reliable way to tell the two apart based on a state machine diagram alone. They would have to query the model to find the block that owns the state machine and look at its operations and receptions to know for sure.

8.5.2.3 Time Events

A **time event** is quite intuitive; it represents an instant in time. When that moment comes during system operation, we say that the time event occurred. And that occurrence, of course, may trigger a transition in a state machine. It's also possible that a time event can occur multiple times during system operation. If so, each occurrence can potentially trigger another transition between states.

There are two types of time events: relative and absolute. In contrast to the ambiguity in the notations for call event triggers and signal event triggers, it's easy to identify a time event trigger. A **relative** time event trigger always begins with the keyword *after*. An **absolute** time event trigger always begins with the keyword *at*. Both types are then followed by a time expression in parentheses.

Time expressions for absolute time events can be as specific or as general as you need them to be—for example, *at (3:00 a.m. GMT)*, *at (Monday)*, *at (3:00 a.m. GMT, Monday, March 4, 2013)*. Remember, however, that a time event will occur multiple times if your time expression is general enough to allow it. The time expression *at (3:00 a.m. GMT)* will result in a new event occurrence every time that moment arrives. And each occurrence of that time event can trigger another transition in a state machine.

Time expressions for relative time events are written as time durations—for example, *after (1 min)*, *after (50 ns)*, *after (1 month)*. When that specified amount of time elapses, the relative time event occurs, and that occurrence can trigger a transition in a state machine.

Figure 8.9 displays an excerpt of the *Attitude Control* state machine in Figure 8.1. This excerpt shows an example of a relative time event trigger on the transition from the *No Comm Link* substate (within the *On-Station* state) to the *Safe Mode* state. The relative time event counter starts when the state machine transitions from the *Have Comm Link* state to the *No Comm Link* state. If two minutes elapse while the system is in the *No Comm Link* state, that relative time event occurs, and the outgoing transition to the *Safe Mode* state fires.

Of course, that may not happen. While the system is in the *No Comm Link* state, it's possible that the *commLinkRestored* signal event will occur before two minutes elapse. If so, the transition back to the *Have Comm Link* state will fire. If the state machine later returns to the *No Comm Link* state, the relative time event counter will be reset upon reentry into that state.

An **external self-transition** (an external transition drawn from a state back to itself) counts as exiting and reentering that state; it will

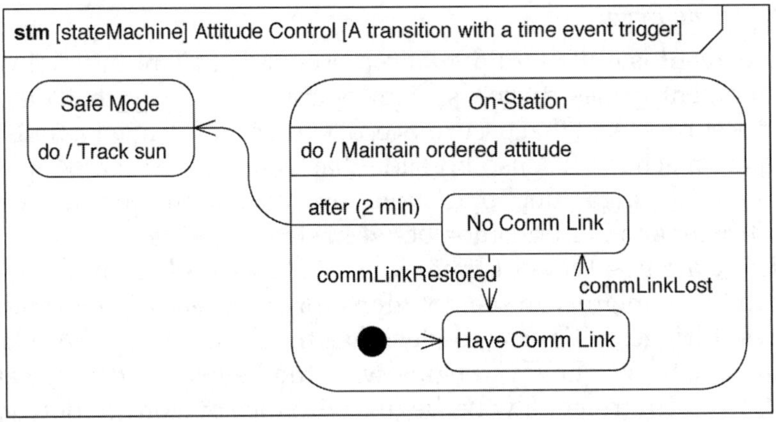

Figure 8.9 *A time event trigger in a state machine*

therefore reset any relative time event counter for that state. In contrast, an **internal transition** within a state does not result in exiting and reentering that state; it will *not* reset a relative time event counter for that state.

Also remember that, like event occurrences of all types, a time event occurrence could potentially interrupt a *do* behavior in a state that has one. If the *No Comm Link* state had a *do* behavior and its execution lasted for two minutes or more, then that execution would be interrupted when the relative time event occurrence caused the transition to the *Safe Mode* state to fire. Be cautious in your daily practice about inadvertently modeling race conditions.

A state's *entry* behavior, in contrast, is regarded as atomic and instantaneous; it's part of the run-to-completion step that constitutes the transition between two states. A relative time event counter starts only when the run-to-completion step finishes executing. An *entry* behavior, therefore, will not be interrupted by a time event occurrence.

8.5.2.4 Change Events

A **change event** is defined as a Boolean expression. A defined change event occurs during system operation each time the specified Boolean expression toggles from false to true. As with a time event trigger, it's easy to identify a change event trigger; it always begins with the keyword *when*, followed by a Boolean expression in parentheses. That Boolean expression is allowed to contain arguments passed into the state machine via other event occurrences, as well as properties owned

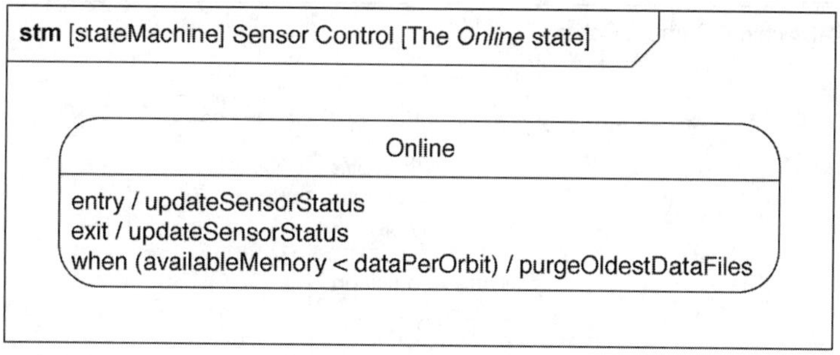

Figure 8.10 *A change event trigger in a state machine*

by the state machine itself and properties owned by the block that executes the state machine.

Figure 8.10 displays an excerpt of the *Sensor Control* state machine in Figure 8.6, this time with a focus on the *Online* state. This state has an internal transition with a change event trigger, *when (availableMemory < dataPerOrbit)*. This change event occurs every time the specified Boolean expression toggles from false to true. And each time it does, the transition effect—the *purgeOldestDataFiles* behavior—gets executed. (To be clear, a change event trigger can also appear on external transitions; it's not limited to internal transitions.)

8.6 Pseudostates

In Section 8.5, I indicate that a state machine can contain two kinds of vertices: *states* and *pseudostates*. The difference between them is straightforward. A state machine can rest in a state; it cannot rest in a pseudostate. You add **pseudostates** to a state machine to impose control logic on the transitions among states.

SysML defines nine kinds of pseudostates. Most of the time, however, you need only two kinds: initial pseudostates and junction pseudostates. Figure 8.11 shows examples of both kinds.

An **initial pseudostate** simply indicates the first state that the state machine will be in when it begins executing (or the first substate that a composite state will be in when it becomes active). The notation for an initial pseudostate is a small, filled-in circle. It's not allowed to have an incoming transition, and the outgoing transition is not allowed to have a trigger or a guard. If it did, that would imply that the state machine might have to rest in the initial pseudostate while waiting for an event

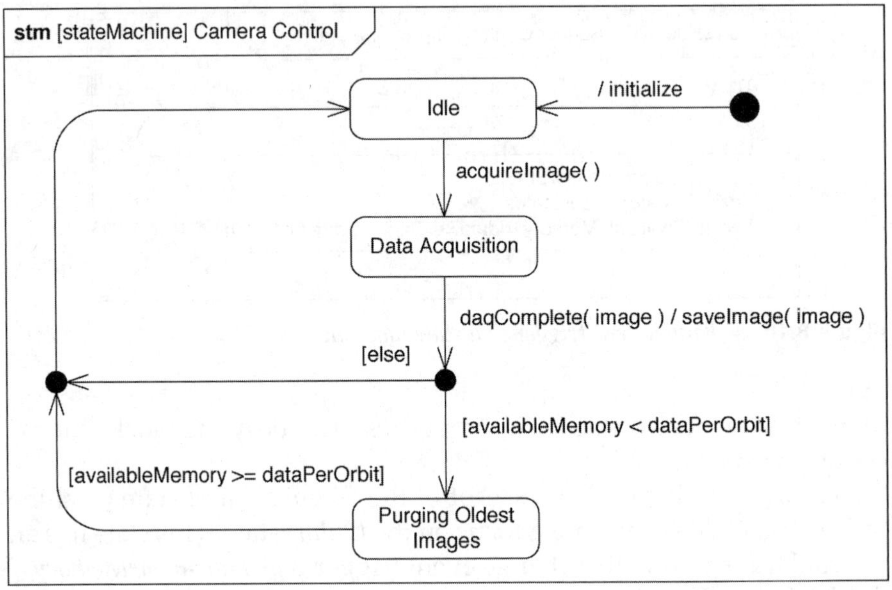

Figure 8.11 *Pseudostates in a state machine*

to occur or for the guard to become true . . . and state machines are not allowed to rest in pseudostates. The outgoing transition can, however, have an effect.

Figure 8.11 conveys that when the *Camera Control* state machine gets invoked, the *initialize* behavior will get executed and the state machine will start in the *Idle* state. Recall that a transition effect is part of the run-to-completion step, which is regarded as atomic and instantaneous. The state machine therefore starts immediately in the *Idle* state; there is no passage of time during the execution of the *initialize* behavior.

The other kind of pseudostate that you will commonly use is a junction pseudostate. A **junction pseudostate** enables you to combine multiple transitions between states into a single (more readable) compound transition. The notation for a junction pseudostate is also a small, filled-in circle—often smaller than an initial pseudostate (though this depends on the modeling tool you're using). Unlike an initial pseudostate, a junction pseudostate must have one or more incoming transitions and one or more outgoing transitions. A single junction pseudostate may, in fact, have multiple incoming transitions *and* multiple outgoing transitions simultaneously.

A junction pseudostate with multiple incoming transitions serves to merge transitions from multiple source vertices, all of which have a common target vertex. The junction pseudostate that leads to the *Idle* state in Figure 8.11 is an example.

A junction pseudostate with multiple outgoing transitions serves as a decision point that leads to one of several alternative target vertices. In this case, each of the outgoing transitions must have a guard specified. The transition whose guard evaluates to true when the trigger event occurs is the one that will fire. If all of the guards happen to be false at that moment, the state machine will remain in its current state. You can optionally specify the guard *else* on (at most) one transition to guarantee that a transition will fire each time the decision is made.

The state machine in Figure 8.11 displays a junction pseudostate (immediately following the *Data Acquisition* state) that serves as a decision point. Because one outgoing transition has the guard "else," the compound transition out of the *Data Acquisition* state is guaranteed to fire each time the *daqComplete* signal event occurs. And when it does, the *saveImage* behavior will get executed as part of the run-to-completion step.

Note that you're allowed to chain several junction pseudostates together to construct an arbitrarily complex compound transition between states. And each segment along that path can have its own guard and effect. However, I recommend that you avoid doing this if possible. The rules for the guards and effects on compound transitions are a bit nuanced, and a diagram is only as good as your stakeholders' ability to interpret it.

8.7 Regions

Like an activity and an interaction, a state machine can convey concurrent behaviors. You do this by adding multiple **regions** to the state machine. Each region contains its own set of vertices and transitions. And each region responds to event occurrences independently of the others. For this reason, we describe regions as **orthogonal** to each other.

You convey that a state machine has multiple regions by using dashed lines that divide the contents area of the state machine diagram, as shown in Figure 8.12. (A state machine with only one region—the most common case—wouldn't display any dashed lines.) You can optionally specify a name for each region; that name would appear as a label floating somewhere inside the region, typically off in a corner.

Everything I state about transitions, states, and pseudostates earlier in the chapter also applies when a state machine has multiple regions. What's new is the key idea that each region must have exactly one active state at any given moment during system operation. To state this differently, a state machine is concurrently in multiple states (one state per region) during an execution of the state machine behavior.

This key idea has implications for the way you construct your state machine diagram. Specifically, it is illegal to draw a transition between two vertices that are in different regions. If such a transition existed, it's possible that a region would be left without an active state at some point—and that's not permissible.

As stated earlier, each region responds to event occurrences independently of the others. To put this more concretely, an event occurrence may cause a transition to fire in one region, but not the others. It's also possible for a single event occurrence to cause multiple transitions to fire, but at most one transition per region. This could happen if two or more regions contained transitions with the same event trigger.

The state machine shown in Figure 8.12 has two orthogonal regions. When the *Telemetry Stream Control* state machine executes, it will be in the *Streaming* or *Suspended* state and concurrently in the *Unencrypted* or *Encrypted* state. These two regions will respond to event occurrences independently of each other. In this particular state machine, the or-

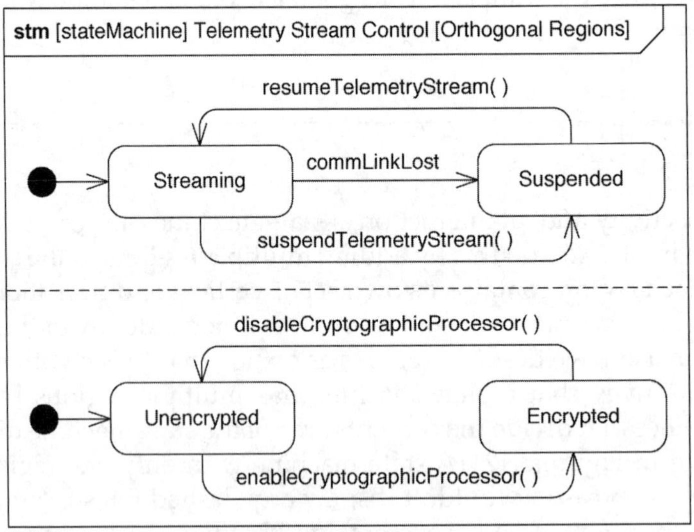

Figure 8.12 *A state machine with orthogonal regions*

thogonal regions have no transition triggers in common. Therefore, only one region will have a transition that fires each time an event occurs.

Summary

A state machine diagram enables you to express information about a system's state-based behavior in response to event occurrences. Not all structures within a system have a defined set of states with transitions among those states. For those that do, the state machine diagram is a uniquely capable medium for communicating the behaviors of those structures to your stakeholders.

Chapter 9

Parametric Diagrams

The parametric diagram is a unique kind of SysML diagram, one that's used to express information about a system's constraints. These constraints generally take the form of mathematical models that determine the set of valid values within a running system. A parametric diagram is uniquely capable of conveying these mathematical models to stakeholders.

9.1 Purpose

SysML allows you to model equations and inequalities as constraint blocks. In Chapter 3, "Block Definition Diagrams," you learned that a **constraint block** is a special kind of block that encapsulates a constraint expression: the equation or inequality you need to model.

Capturing a constraint expression in your SysML model, however, is simply a means to an end. The power of this capability emerges when you apply the constraint expression to a block somewhere in your model to impose a fixed mathematical relationship on that block's value properties.

What do you achieve by imposing a fixed mathematical relationship on a block's value properties? You gain these abilities:

- To specify assertions about valid system values within an operational system (and therefore detect exceptional conditions when they occur)

- To use the blocks in your system model to provide the inputs for (and capture the outputs of) engineering analyses and simulations during the design stage

These advanced modeling practices are among the most powerful capabilities SysML offers. It's worth your time to become comfortable with the practice of capturing constraint expressions in your model and applying them to blocks.

You apply a constraint expression to a block by **binding** each variable in the expression (formally called a **constraint parameter**) to a value property that exists somewhere in your model. That value property may belong to the block itself, or it may belong to one of the block's part properties or reference properties. In this way, SysML allows you to create an arbitrarily complex mathematical model and then "wire" it (colloquially speaking) to various parts of an arbitrarily complex structural model.

Where do **parametric diagrams** fit into all this? They serve two purposes:

- To display the bindings between constraint parameters *in different constraint expressions* to create a composite system of equations (or inequalities)
- To display the bindings between constraint parameters and value properties to apply a constraint expression to a block (and, in so doing, impose a fixed mathematical relationship on a set of value properties)

In this chapter, I discuss in detail how to create parametric diagrams that fulfill these purposes. I provide concrete examples of parametric diagrams as well as their corresponding BDDs to help you more easily understand the relationships between the kinds of elements that appear on each kind of diagram.

9.2 When Should You Create a Parametric Diagram?

SysML defines a parametric diagram as a special kind of internal block diagram (IBD). Like an IBD, a parametric diagram displays the internal structure of a block—but with a focus on the bindings between value properties and constraint parameters. Just as IBDs and BDDs provide complementary views of blocks, so do parametric diagrams and BDDs.

Because of this close relationship, you can potentially create parametric diagrams during any stage of the system life cycle. I say "potentially" because not all modeling teams will necessarily create a SysML parametric model. This is an advanced modeling practice; it requires a high degree of fidelity in your system model, something that has an associated cost. You would incur the cost of creating a parametric model (and displaying it on parametric diagrams) only if it's within the model scope required to fulfill the model's purpose as defined in your team's project plan.

9.3 Blocks, Revisited

To fully make sense of what you will see on the parametric diagrams that appear later in this chapter, it's helpful first to see the corresponding BDDs. A BDD is the kind of diagram you create to display the *definitions* of blocks and constraint blocks. A parametric diagram then displays the *usages* of those blocks and constraint blocks, with a focus on the bindings between value properties and constraint parameters. The BDDs in Figures 9.1 and 9.2 display the blocks and constraint blocks that are needed to construct corresponding parametric diagrams.

The BDD in Figure 9.1 is a repeat of the one shown in Figure 3.26 in Chapter 3. This BDD displays three constraint blocks: *Hohmann Transfer*, *Transfer Orbit Size*, and *Transfer Time of Flight*. The *Hohmann Transfer* constraint block is composed of the other two; it owns one constraint property named *ttof* of type *Transfer Time of Flight* and another constraint property named *tos* of type *Transfer Orbit Size*. These constraint properties are displayed on this diagram in two ways:

- Name strings displayed in the constraints compartment of *Hohmann Transfer*
- Role names on the part ends of the composite association relationships

Transfer Orbit Size and *Transfer Time of Flight* each encapsulate a constraint expression. And each of those constraint expressions contains three constraint parameters, which are displayed as name strings in the parameters compartments. These constraint parameters (like most constraint parameters) are typed by value types that exist somewhere in the system model.

The BDD in Figure 9.1 conveys that the *Hohmann Transfer* constraint block defines a constraint expression that is a composite of the two

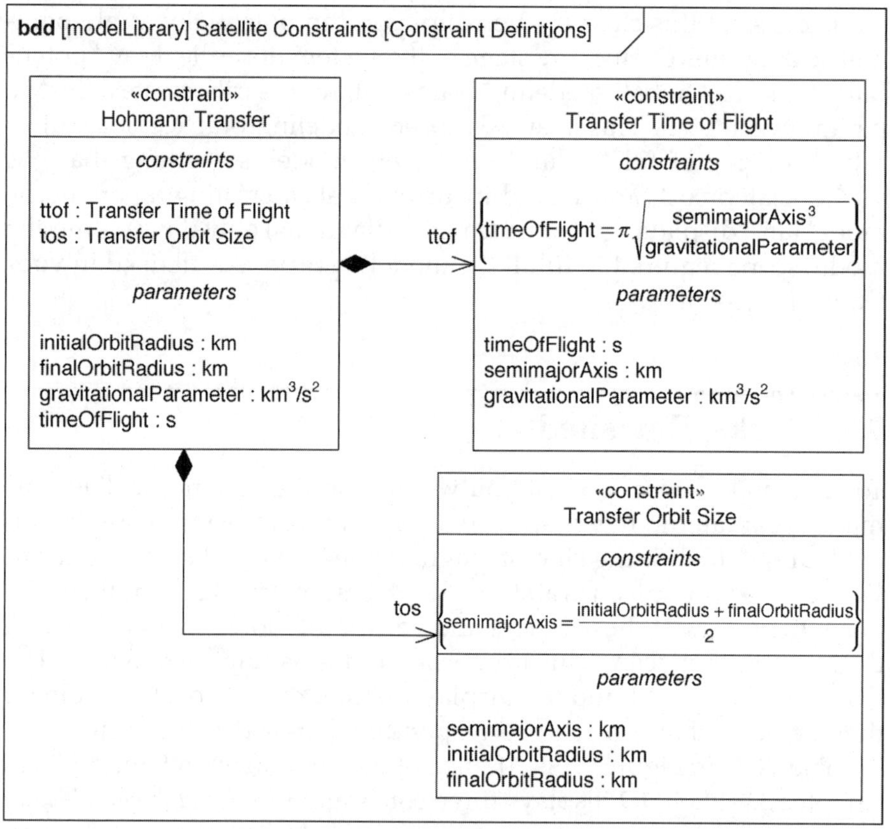

Figure 9.1 *Block definitions needed to create a parametric diagram for* Hohmann Transfer

simpler ones. But this diagram doesn't convey *where* those two simpler constraint expressions are specifically connected to each other to create the composite constraint expression; that's what a parametric diagram is for. The parametric diagram in Figure 9.3 displays this complementary piece of information.

The BDD in Figure 9.2 displays the subset of blocks that are needed to create a parametric diagram for a transfer time analysis (shown later in Figure 9.4). I've employed the common and useful technique of creating a block, *Transfer Time Analysis*, to represent the analysis context itself. (Some modeling teams even define the custom stereotype «analysisContext» to distinguish this block from the others in the model; please note that this stereotype is not defined in SysML.)

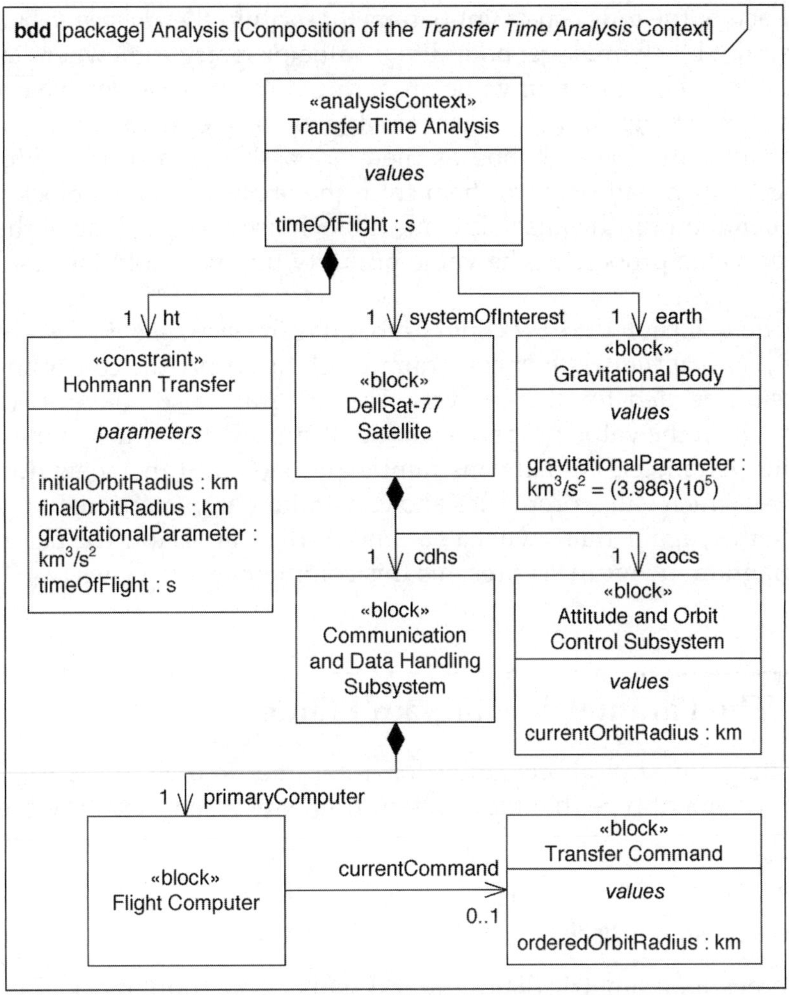

Figure 9.2 *Block definitions needed to create a parametric diagram for* Transfer Time Analysis

By convention, the analysis context block has composite associations with one or more constraint blocks that represent the types of the constraint properties that are needed for the analysis. Also by convention, the analysis context block has reference associations with one or more blocks that own the value properties that will supply the values needed to perform the analysis.

Each of the four constraint parameters within the *Hohmann Transfer* constraint block must get bound to a value property somewhere in the system model. Those four value properties do not necessarily have to be owned by a single block; it's permissible for each of them to be owned by a different block, as long as there are association relationships in place to form pathways to them from the analysis context block. (It's also common for the analysis context block itself to own one of the required value properties: the value property that will hold the result of the analysis.)

The BDD in Figure 9.2 conveys that the *Transfer Time Analysis* block uses the *Hohmann Transfer* constraint block to supply the constraint expression needed for the analysis. This diagram also conveys which blocks own the value properties that will supply the values to the constraint parameters in that constraint expression. But this view doesn't convey which value properties and constraint parameters are bound to each other; again, that's what a parametric diagram is for. The parametric diagram in Figure 9.4 presents this complementary view.

9.4 The Parametric Diagram Frame

The diagram kind abbreviation for a parametric diagram is *par*. The model element type that the diagram frame represents can be either of these:

- *block*
- *constraintBlock*

When a parametric diagram represents a constraint block, the diagram displays only the constraint properties and bindings that form the internal structure of that constraint block. A parametric diagram that represents a constraint block is shown in Figure 9.3.

When a parametric diagram represents a block, it primarily displays the bindings between the block's value properties and constraint properties. But it may also show that block's part properties and reference properties if they contain nested value properties of interest. A parametric diagram that represents a block is shown in Figure 9.4.

The diagram header in Figure 9.3 tells us that the frame of this parametric diagram represents the *Hohmann Transfer* constraint block (which appears on the BDDs in Figures 9.1 and 9.2). The name of this diagram

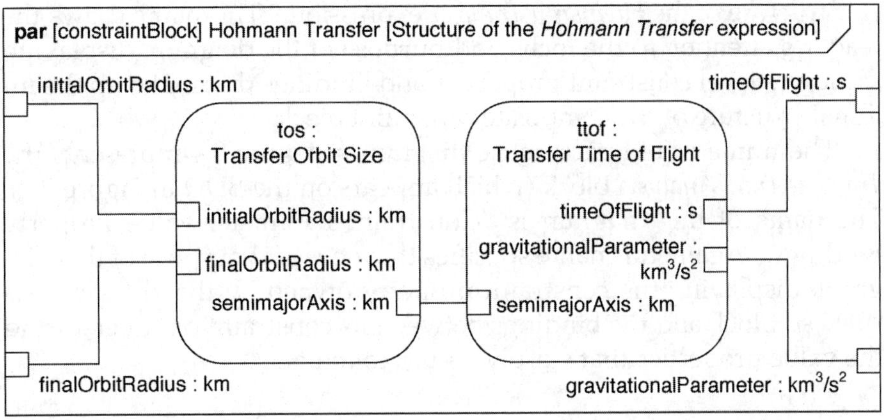

Figure 9.3 *Parametric diagram of the* Hohmann Transfer *constraint block*

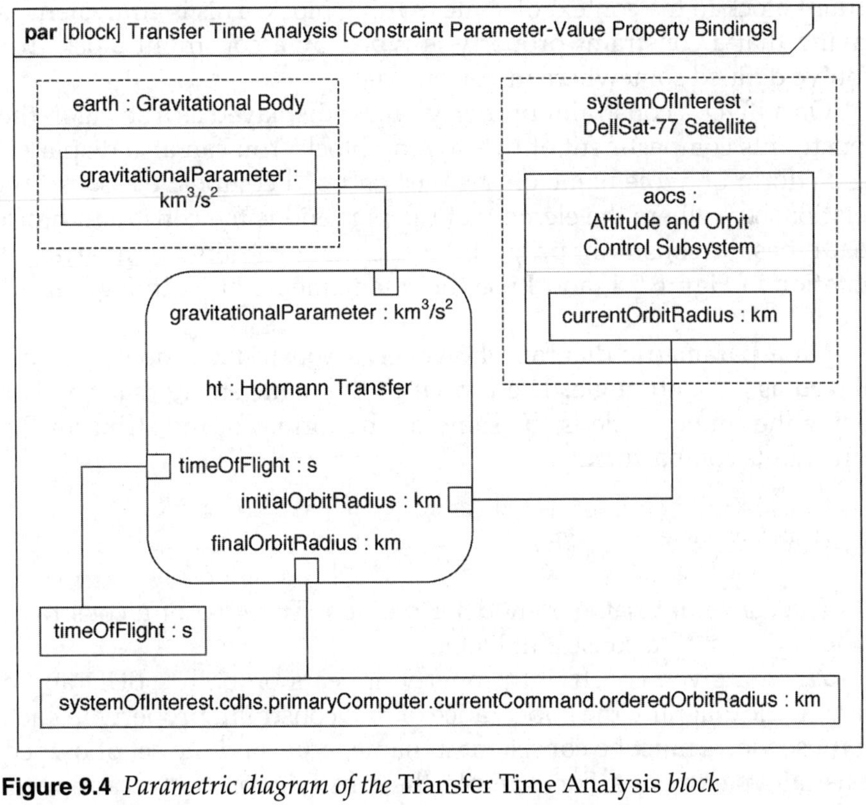

Figure 9.4 *Parametric diagram of the* Transfer Time Analysis *block*

is "Structure of the *Hohmann Transfer* expression." This name draws the reader's attention to the focus and purpose of the diagram: displaying the lower-level constraint properties and bindings that make up the internal structure of the composite constraint block.

The frame of the parametric diagram in Figure 9.4 represents the *Transfer Time Analysis* block (which appears on the BDD in Figure 9.2). The name of this diagram is "Constraint Parameter-Value Property Bindings." Again, this name specifies the focus and purpose of this diagram: displaying the constraint property owned by the *Transfer Time Analysis* block and the bindings between its constraint parameters and the value properties that supply values to them.

9.5 Constraint Properties

In Chapter 3 you learned that a constraint property is a usage of a constraint block in the context of some owning block. This is equivalent to saying that a constraint property is typed by a constraint block that you've defined somewhere in your model.

On a BDD, a constraint property can be displayed as a string in the constraints compartment of the owning block. You can also display it in the form of a role name on the part end of a composite association relationship (where the element at the part end is the constraint block that types the constraint property). I use the constraints compartment notation in Figure 9.1, and I use the role name notation in Figures 9.1 and 9.2.

On a parametric diagram, however, a constraint property is displayed as a round-angle. The format of the name string that appears inside the round-angle is the same as the format of the string in the constraints compartment:

```
<constraint name> : <type>
```

The constraint name is modeler defined. The type of a constraint property must be a constraint block.

When a given constraint property appears on both a BDD and a parametric diagram, the appearance of that constraint property in each of those views must be consistent in name, type, and the set of owned constraint parameters. For example, the constraint property *tos* appears on the BDD in Figure 9.1 and on the parametric diagram in Figure 9.3. Both views of *tos* convey that it is a usage of (that is, typed by) the

Transfer Orbit Size constraint block. Both views also convey that *tos* owns three constraint parameters. The constraint property *ht*, which appears on the BDD in Figure 9.2 and on the parametric diagram in Figure 9.4, provides another example of this correspondence between views.

9.6 Constraint Parameters

As mentioned in Chapter 3, **constraint parameter** is the formal term for a variable that appears in a constraint expression. On a BDD, a constraint parameter can be displayed as a string in the parameters compartment of the owning constraint block. On a parametric diagram, a constraint parameter is displayed as a small square attached to the boundary on the inside of a constraint property. A constraint parameter can also be attached to the frame of a parametric diagram when the diagram represents a constraint block.

The format of the name string for a constraint parameter is the same on both a BDD and a parametric diagram:

```
<parameter name> : <type> [<multiplicity>]
```

The parameter name is modeler defined. The type is almost always a value type that you've defined somewhere in your model. (SysML allows for a constraint parameter to be typed by a block, but the meaning of this is poorly defined, and modelers rarely do it in daily practice.) The multiplicity represents a constraint on the number of values of the specified type that the constraint parameter can hold. The default multiplicity, if not shown, is 1..1.

I mention in Section 9.5, "Constraint Properties," that the appearances of a constraint property on a BDD and on a parametric diagram must be consistent. Similarly, the appearance of the name strings for the constraint parameters must be consistent in both of these views of the model.

9.7 Value Properties

A **value property** is a usage of a value type in the context of an owning block. This is equivalent to saying that a value property is typed by a value type that you've defined somewhere in your model. A value property could represent a quantitative characteristic of a block as well as a Boolean value or a string. Value properties matter in the context of a parametric model, because they supply values to constraint parameters so that you (or an equation-solving tool) can evaluate constraint expressions.

On a BDD, a value property is displayed as a string in the values compartment of the owning block. On a parametric diagram (one that represents a block), a value property is displayed as a rectangle with a solid boundary. (You wouldn't display a value property on a parametric diagram that represents a constraint block, because constraint blocks can't own value properties.)

The format of the name string that appears inside the rectangle on the parametric diagram is the same as the format of the string in the values compartment:

```
<value name> : <type> [<multiplicity>] = <default value>
```

The value name is modeler defined. The type, again, must be a value type that you've defined somewhere in your model. The multiplicity means the same thing that it does in Section 9.6, "Constraint Parameters." The default value is an optional piece of information; when shown, it represents the specific value assigned to the value property before any other value gets assigned during system operation (or at analysis time).

When a value property is owned by the block that's named in the parametric diagram header, the rectangle notation for that value property floats somewhere inside the diagram frame. Stated more precisely, the diagram frame itself is the boundary that encapsulates the value property. For example, the BDD in Figure 9.2 conveys that the *Transfer Time Analysis* block owns a value property named *timeOfFlight*. The parametric diagram in Figure 9.4 represents the *Transfer Time Analysis* block. The *timeOfFlight* value property therefore appears nested within the diagram frame; it's not encapsulated by any other boundary.

The value properties that supply values to constraint parameters do not necessarily have to be owned by the block that's named in the parametric diagram header; they may be owned instead by that block's part properties and reference properties. In fact, the value properties of interest may be deeply nested within the block's structure.

You can display a deeply nested value property on a parametric diagram by using either dot notation or nesting notation. (I discuss both notations in detail in Chapter 4, "Internal Block Diagrams," in the context of IBDs.) On the parametric diagram shown in Figure 9.4, I use dot notation to display the *orderedOrbitRadius* value property, and I use nesting notation to display the *gravitationalParameter* and *currentOrbitRadius* value properties.

When you use dot notation, the rectangle for the value property is nested within the diagram frame; it does not appear encapsulated by any other boundary. However, the string within the rectangle conveys that the value property is, in fact, deeply nested. The dot notation string contains the names of all the part properties (and reference properties) in the hierarchy, down to the name of the value property itself. The type of the value property appears after a colon at the end of the string.

Note

Be aware that dot notation allows you to display only the *names* of the part properties and reference properties in the hierarchy and not their *types*. If your target audience needs to see that information on the parametric diagram, then you should use nesting notation instead.

When you use nesting notation, the value property appears encapsulated by the boundary of the part property or reference property that owns it. Recall from Chapter 4 that a part property is displayed as a rectangle with a solid boundary; a reference property is displayed as a rectangle with a dashed boundary.

The BDD in Figure 9.2 conveys that the *Transfer Time Analysis* block has a reference property named *systemOfInterest* (of type *DellSat-77 Satellite*), which in turn owns a part property named *aocs* (of type *Attitude and Orbit Control Subsystem*), which in turn owns a value property named *currentOrbitRadius* (of type *km*). The parametric diagram in Figure 9.4 therefore displays *systemOfInterest* with a dashed boundary and *aocs* with a solid boundary. The value property *currentOrbitRadius* appears nested within its owner, *aocs*.

You likely noticed that the notations for a value property and a part property are the same: a rectangle with a solid boundary. This may strike you as ambiguous. However, there is a way to definitively distinguish one from the other: Value properties get bound to constraint parameters, and part properties do not.

9.8 Binding Connectors

I've used the terms *bound* and *binding* repeatedly throughout this chapter. This word choice is not arbitrary. SysML defines a special kind of connector called a **binding connector**, which simply represents an equality relationship between the two elements attached at either end. One of those two bound elements must be a constraint parameter. The other bound element can be either a value property or another constraint parameter (in a different constraint expression).

A binding connector can appear only on a parametric diagram. The notation for a binding connector is a solid line attached to the boundaries of the two bound elements. All the connectors on the parametric diagrams in Figures 9.3 and 9.4 are binding connectors.

Note that binding connectors convey no notion of direction. When a value is assigned to a value property—during either system operation or analysis execution time—the constraint parameter at the other end of the binding connector instantaneously assumes the same value. That value is then available to the constraint property that owns the constraint parameter, and one of two things can happen:

- The constraint property evaluates to true or false (if values are supplied to *all* of the parameters in the expression).
- A value is calculated for the constraint parameter that didn't receive a value across its binding connector (if values are supplied to all of the others).

This implies a more general point: Constraint properties are noncausal; no constraint parameter in the expression is assumed ahead of time to be the *dependent* variable. And this is true even if the constraint expression is written with a variable isolated on one side. The dependent variable can change in each analysis run based on which value properties receive assignments.

Summary

A parametric diagram expresses a set of constraints—generally, equations and inequalities—that determine the values that are valid in a system that's operating nominally. A parametric diagram is the only one of the nine kinds of SysML diagrams that can convey this aspect of a system's design. Not all modeling teams need to create a mathematical model of a system to fulfill the model's purpose defined in the project plan. For those that do, the parametric diagram is an essential medium for communicating this kind of information to stakeholders.

Chapter 10

Package Diagrams

You can display various kinds of elements and relationships on a package diagram to express information about the structure of a system model. To correctly interpret (and create) package diagrams, you need to understand the concept of namespace containment as well as the various notations that you can use to convey namespace containment on diagrams. There are four specialized kinds of packages, each with its own purpose, and certain kinds of dependencies can exist among them. You need to know the similarities and differences between a package diagram and a BDD to determine which one is best depending on where you want the focus to be.

10.1 Purpose

A **package diagram** is the kind of diagram you create when you need to display the organization of the system model. The organization of the system model is determined by the package hierarchy you create to partition the elements in the model into logically cohesive groups. There is no one right way to structure your system model. Different methodologies propose various model structures, and they will be more or less effective for your team depending on your project's unique goals.

Once you decide on an effective model structure for your project—and it may take several iterations—it's useful to create package diagrams to provide your stakeholders an easily understood view of that structure. Package diagrams can display packages nested within packages to convey the containment hierarchy of the model.

10.2 When Should You Create a Package Diagram?

Package diagrams convey information about the structure of the model itself. You would therefore create new package diagrams when you modify the model structure in significant ways (as determined by the concerns of your stakeholders). You'll create many new packages (and many new package diagrams) at the beginning of your project when you first create the model structure. And, of course, the design of the model structure—like the design of your system's structure—will evolve over time.

As the design stage of the life cycle progresses, you will decompose higher-level structural elements into lower-level structural elements. It's typical to add new packages to the model structure to contain those new, lower-level structural elements. When you do, you will also create new package diagrams at that time.

10.3 The Package Diagram Frame

The diagram kind abbreviation for a package diagram is *pkg*. The model element type that the diagram frame represents can be any of the following:

- *package*
- *model*
- *modelLibrary*
- *view*
- *profile*

As I discussed in Section 2.4, "General Diagram Concepts," the model element that's named in the diagram header serves as the namespace for the elements shown on the diagram. I mention in Chapter 3, "Block Definition Diagrams," that a namespace is simply a model element that's allowed to contain other named elements. To state this differently, a namespace can have named elements nested under it within the model hierarchy. A namespace, therefore, is a concept that has

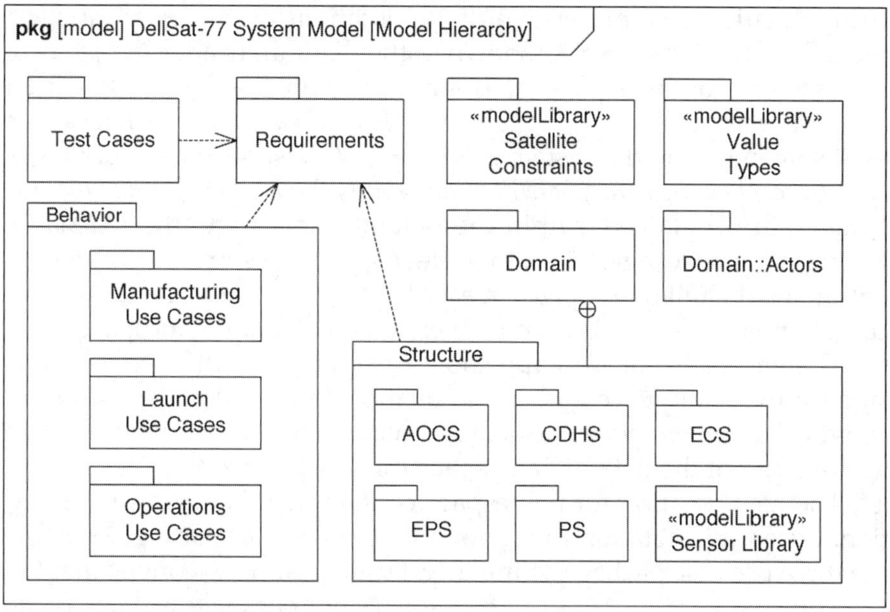

Figure 10.1 *A sample package diagram*

meaning only within your system model and not within an instance of your system.

The package diagram shown in Figure 10.1 represents the model named *DellSat-77 System Model*. A **model** is one of four specialized kinds of packages; it's the kind of package that serves as the root of a containment hierarchy. Therefore, it's also correct to say that the package diagram in Figure 10.1 represents the *package* named *DellSat-77 System Model*. And this package serves as the default namespace for the elements shown on the diagram.

10.4 Notations for Namespace Containment

That term *default namespace* merits elaboration. It's correct to conclude that the package named in the diagram header is the owner (the container) of the elements shown in the contents area—unless another namespace containment relationship is explicitly shown on the diagram. SysML provides three notations to explicitly convey namespace containment: crosshair notation, nesting notation, and qualified name string notation. The presence of these notations on the diagram overrides the default namespace shown in the diagram header.

The notation for a package is a folder symbol—a rectangle with a tab on the upper-left side. Figure 10.1 shows seventeen packages that exist somewhere in the system model. Six of those seventeen packages (*Test Cases*, *Requirements*, *Satellite Constraints*, *Value Types*, *Behavior*, and *Domain*) are contained within the default namespace, the *DellSat-77 System Model* package. The other eleven packages are not contained within the default namespace; in each case, one of the three namespace containment notations is used to override the default namespace.

The crosshair notation appears as a solid line with a circled plus sign on the namespace end of the relationship. I use this notation to convey that the *Structure* package is contained within the *Domain* package (instead of the default namespace for this diagram).

The next notation for namespace containment is the nesting notation. I used this notation to convey that the *Behavior* package contains the three use case packages. Similarly, I used nesting to convey that the *Structure* package contains the five subsystem packages as well as the *Sensor Library* package.

Note

I could have used the crosshair notation again to express these same relationships. The decision to use the crosshair notation or the nesting notation is purely a stylistic choice. Note, however, that when you choose to use the nesting notation, SysML requires that you display the name of the parent package in the tab of the folder symbol, as shown in Figure 10.1.

The final notation for namespace containment—the **qualified name string notation**—is the most space-efficient notation. The string *Domain::Actors* in Figure 10.1 is an example of a qualified name string. The element being named—the *Actors* package—is at the end of the string. The double colon in the string conveys that the *Actors* package is contained within the *Domain* package.

A qualified name string can, of course, be arbitrarily long to specify multiple levels of containment down to an element that is deeply nested within the system model. And even though I am discussing qualified name strings in the context of packages, note that you can use a qualified name string for any element to precisely specify its location in the model hierarchy, no matter which diagrams it appears on.

If a qualified name string begins with the name of the model itself (e.g., *DellSat-77 System Model::Domain::Actors*), we refer to it as a **fully qualified** name. If a qualified name string does not begin with the name of the model (e.g., *Domain::Actors*), we refer to it as a **relative qualified** name; the path down to the element named at the end of the string is relative to (begins with) the element named in the diagram header. Based on this, you can conclude from Figure 10.1 that the *Actors* package is contained within the *Domain* package, which is, in turn, contained within the *DellSat-77 System Model* package.

I could have used the crosshair notation or the nesting notation to express the relationship between the *Domain* package and the *Actors* package. Modelers generally use the qualified name string notation only when the element's owner is not shown on the same diagram.

10.5 Dependencies between Packages

Figure 10.1 shows three dependencies—the dashed lines with open arrowheads—drawn between packages. I discuss dependencies in detail in Section 3.7, "Dependencies," in the context of BDDs. Dependencies mean the same thing here: A change in the supplier element (at the arrowhead end) may result in a change to the client element (at the tail end).

The package diagram shown in Figure 10.1 conveys that a change to the contents of the *Requirements* package may result in a change to the contents of the *Test Cases, Behavior,* and *Structure* packages. Displaying the full set of dependencies among the packages shown on this diagram would make the diagram unreadable. I chose to display these three dependencies to demonstrate that you can use this relationship to relate packages, too. SysML, in fact, imposes no constraints on the kinds of elements that can appear at either end of a dependency.

10.6 Importing Packages

When you create your model hierarchy, you will likely import packages (and their contents) into your system model from other models and packages. You might do this to reuse model libraries that you created earlier in other models. Or you might do it to vertically or horizontally integrate a set of models that you've decentralized (for example,

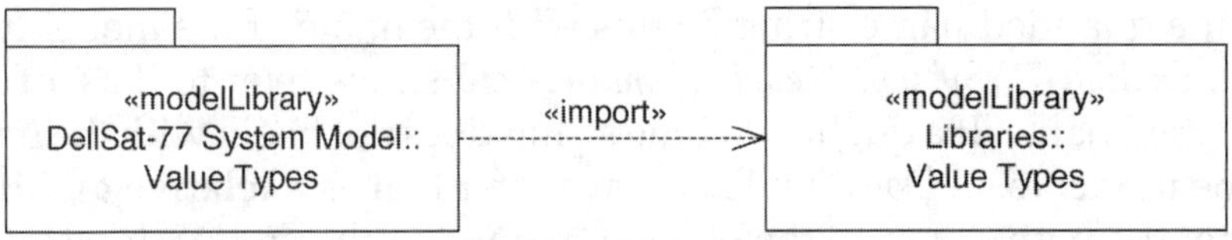

Figure 10.2 *Package import relationship*

integrating a *system* architectural model with a set of *subsystem* architectural models).

SysML provides a mechanism on package diagrams to convey that one package imports the contents of another. The relationship is intuitively called a **package import** relationship. The notation for this relationship is a dashed line with an open arrowhead with the keyword «import» hovering near the line.

An example of this relationship is shown in Figure 10.2. This package diagram fragment conveys that the *Value Types* package in the *DellSat-77 System Model* package imports the contents of the *Value Types* package contained within the *Libraries* package. (The *Libraries* package could be a separate model entirely, although that's not clear from this diagram fragment.)

All commercial-grade SysML modeling tools provide package import functionality, and you will perform this operation often in the early stages of design. With that said, I do not recommend spending your time creating package diagrams to display that you've performed this operation. It's more important to create package diagrams that express the finished model hierarchy than it is to create diagrams that express where each piece of that hierarchy came from.

Next, let's look at specific reasons for importing packages in the context of the various specialized kinds of packages.

10.7 Specialized Packages

As mentioned earlier, a package is simply a container for a set of named elements, some of which may be other packages. It's a mechanism for partitioning the elements in your system model into logically cohesive groups. SysML defines four specialized kinds of packages: model, model library, profile, and view. Each has a distinct purpose beyond its basic function as a container.

10.7.1 Models

A **model** is a kind of package that serves as the root of a containment hierarchy. This is equivalent to saying that it's the top-level package in a hierarchy. The notation for a model is the same as the one for a package—a folder symbol—but a model must have either the keyword «model» preceding the name or a small triangle in the upper-right corner of the folder symbol.

10.7.2 Model Libraries

A **model library** is a kind of package that contains a set of elements you intend to reuse in multiple models. The notation for a model library is the same as the one for a package—a folder symbol—but it must have the keyword «modelLibrary» displayed above the name.

You need not create any particular package as a model library; common examples of model libraries include ones for value types and constraint blocks. You might also create model libraries to contain the definitions (blocks) for low-level hardware and software components that are common to many kinds of systems, such as items on an approved parts list (APL), sensors, operating system libraries, and programming language libraries.

Your project team will cumulatively develop a set of model libraries over time. I recommend that you diligently archive all your model libraries within a separate model dedicated to that purpose (a *Libraries* model). You can then import model libraries as needed from that model into the various system models you create.

10.7.3 Profiles

A **profile** is a kind of package that contains a set of stereotypes. A **stereotype** defines a new kind of model element by adding properties, constraints, or semantics to an existing kind of model element. When you create a profile—a package of stereotypes—you are, in effect, defining a new modeling language that is an extension of an existing modeling language.

You can then **apply** a profile to a package, model, or model library. This means that the package is allowed to contain the new kinds of model elements that exist in the modeling language defined by that profile. The package diagram in Figure 10.3 shows that the model *Reliability Markov Model* applies the *SysML4Modelica* profile. This conveys that *Reliability Markov Model* can contain the new kinds of model elements defined in the *SysML4Modelica* modeling language.

Creating new profiles and stereotypes is an advanced modeling practice that you will not engage in on a daily basis. To gain a deeper understanding of these concepts, I refer you to the excellent book *A Practical Guide to SysML: The Systems Modeling Language* by Sanford Friedenthal, Alan Moore, and Rick Steiner.

10.7.4 Views

A **view** is a kind of package that contains a filtered subset of a model. It is a package that selectively imports other packages, elements, and diagrams in the system model that together represent an aspect of the model that is of interest to a particular set of stakeholders.

In formal terms, a view conforms to a defined viewpoint. A **viewpoint** is a model element that contains five properties: *stakeholders*, *concerns*, *purpose*, *languages*, and *methods*. The *stakeholders* property is a string that lists the stakeholders that would find this viewpoint relevant to their concerns. The *concerns* property is a string that expresses the stakeholder questions that will be answered by the elements and diagrams contained within a conforming view. The *purpose* property is a string that specifies the reason you're defining this viewpoint. The *languages* property is a string that lists the modeling languages that will be used in a conforming view. The *methods* property is a string that specifies the set of rules that you (or the modeling tool) will follow to construct a conforming view.

If you intend to build a conforming view yourself (by manually performing package import operations), then your *methods* string can be a phrase or sentence expressed in an informal natural language (such as English). If you intend to have your SysML modeling tool construct a conforming view in an automated way, then your *methods* string will have to be a set of statements expressed in a formal query language that is supported by that tool.

The package diagram in Figure 10.3 displays a viewpoint, *Reliability Viewpoint*, with values specified for its five properties that together define a particular aspect of the system model that is relevant to the listed stakeholders. This diagram also displays the view (*Reliability View*) that conforms to that viewpoint.

A «conform» relationship is simply a special kind of dependency—a dashed line with an open arrowhead—drawn from the view to the viewpoint. This relationship conveys that *Reliability View* is a view—a

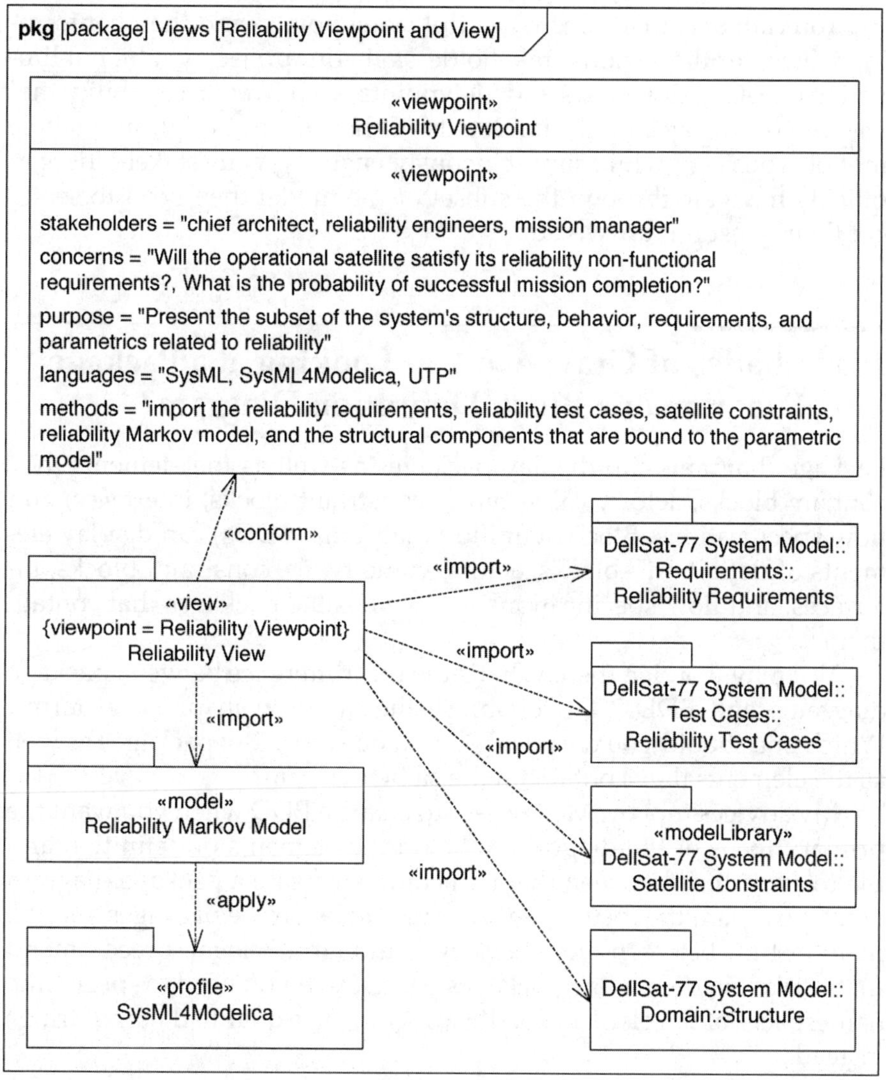

Figure 10.3 *A viewpoint and a conforming view*

special kind of package—that is constructed according to the rules listed in the *methods* property of *Reliability Viewpoint*; "constructed" in this context means that *Reliability View* imports other packages (and their elements and diagrams) to create a filtered subset of the system model that together addresses the listed concerns of the applicable stakeholders.

You can, of course, create multiple viewpoints (and their conforming views) for the various stakeholders on your project; you're not limited to one. Creating views and viewpoints is a powerful capability that allows you to capture all the details of the system design in a single model repository while simultaneously enabling your stakeholders to quickly navigate through the subset of the model they need to see to get their questions answered.

10.8 Shades of Gray: Are You Looking at a Package Diagram or a Block Definition Diagram?

Package diagrams can display packages as well as the elements they contain: blocks, actors, value types, constraint blocks, interfaces, and flow specifications. Block definition diagrams (BDDs) can display elements of definition—blocks, actors, value types, constraint blocks, interfaces, and flow specifications—as well as the packages that contain them.

You may wonder, then: What's the real difference between package diagrams and BDDs? Or to phrase the question in practical terms: Which one should you create when you need to show packages as well as the elements they contain on the same diagram?

My advice—not SysML law—is to create a BDD when you want the primary focus of the diagram to be on the elements of definition and the relationships between them; in contrast, create a package diagram when you want the focus of the diagram to be on the packages and the relationships between *them*. And even though it's legal, I recommend that you avoid displaying packages on BDDs. If you need to specify the namespaces of the elements on the diagram, use qualified name strings instead.

Summary

A package diagram expresses information about the structure of a system model (a package containment hierarchy); this is in contrast to BDDs and IBDs, which convey information about the structure of a system you're designing. Modelers often create package diagrams to convey the logical groupings of model elements and to aid stakeholders in navigating the model structure when they need to locate particular elements.

There are four specialized kinds of packages defined in SysML. A model is the kind of package that serves as the root of a containment hierarchy. A model library is the kind of package that contains a set of elements that you intend to reuse in multiple models. A profile is the kind of package that contains a set of stereotypes—extensions to an existing modeling language that define a new modeling language that's better suited for a particular design domain. A view is the kind of package that contains a filtered subset of a model—a subset that conforms to a defined viewpoint and addresses specific stakeholder concerns. You can create package diagrams to display these specialized kinds of packages and the dependencies that exist among them.

Chapter 11

Requirements Diagrams

Another essential kind of SysML diagram is the requirements diagram. Modelers typically use six kinds of relationships to establish traceability among requirements as well as traceability from requirements to structures and behaviors in the system model. You can use several notations on requirements diagrams to express these relationships, and each has strengths and weaknesses.

A system's requirements inform every other aspect of its design. The requirements diagram is the primary medium in SysML for conveying this kind of information to your stakeholders.

11.1 Purpose

Text-based requirements (and the **requirements specifications** that contain them) have traditionally been staples in systems engineering. This is not to imply that all methodologies require text-based requirements or that your project team must create them. An increasingly widespread technique is to create use cases (and their associated use case narratives) in lieu of text-based functional requirements, and constraint expressions in lieu of text-based nonfunctional requirements. If your project team does write text-based requirements, however, the **requirements diagram** is the kind of SysML diagram you would create when you need to display those requirements and their relationships to other model elements.

This diagram is particularly valuable when your target audience needs to see the traceability from the requirements to the elements in your system model that are dependent on them. SysML imposes no constraints on what those dependent elements may be. The most common kinds include blocks, use cases, test cases, and other requirements.

11.2 When Should You Create a Requirements Diagram?

As you add new elements to the model, you will create relationships from those elements back to the requirements that drove the need for their creation. Establishing requirements traceability in this manner is an ongoing activity throughout design and development. And you may need to create a requirements diagram to display those relationships at any point during this work.

11.3 The Requirements Diagram Frame

The diagram kind abbreviation for a requirements diagram is *req*. The model element type that the diagram frame represents can be any of the following:

- *package*
- *model*
- *modelLibrary*
- *view*
- *requirement*

The model element that's named in the diagram header is the default namespace for the elements shown in the contents area. You know from earlier chapters that a namespace is simply a model element that's allowed to contain other named elements. This means that a namespace can have named elements nested under it in the model hierarchy. As before, a namespace is a concept that has meaning only within your system model and not in an instance of your system.

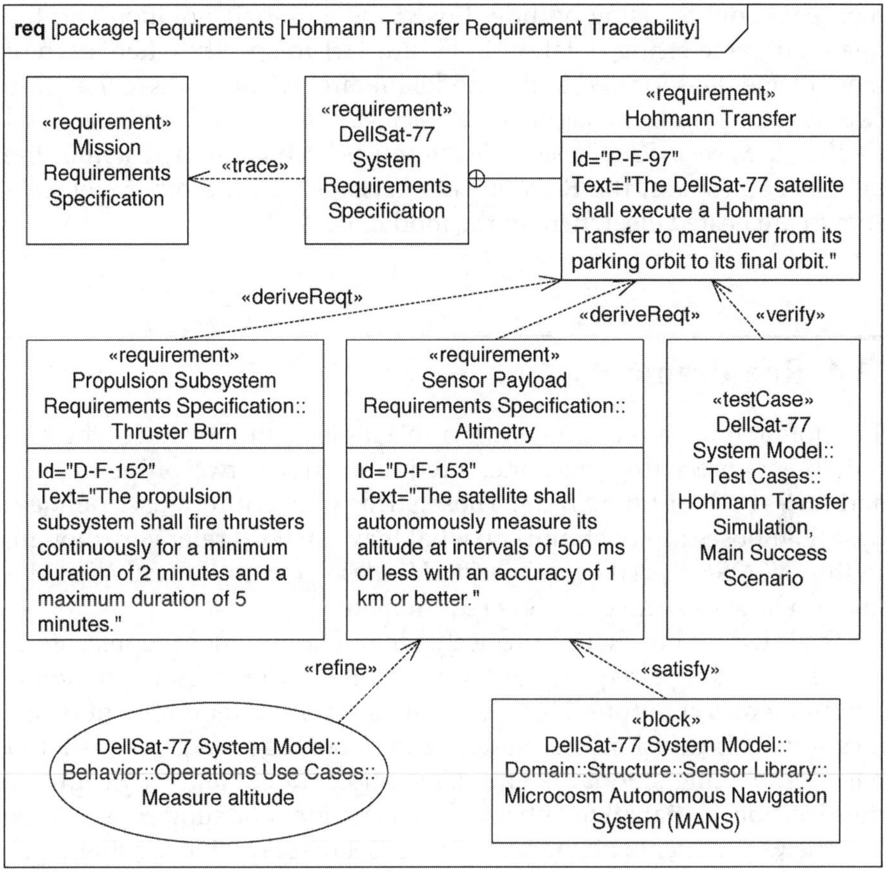

Figure 11.1 *A sample requirements diagram*

The requirements diagram in Figure 11.1 is named "Hohmann Transfer Requirement Traceability." This diagram represents the *Requirements* package that exists somewhere in the system model. The *Requirements* package therefore is the default namespace for the elements shown in the contents area.

Recall from Chapter 10, "Package Diagrams," however, that SysML provides three notations to explicitly convey namespace containment, and these notations (crosshair notation, nesting notation, and qualified name string notation) override the default namespace. Two of these notations—crosshair notation and qualified name string notation—are legal on a requirements diagram as well.

In the sections that follow I discuss in detail the elements shown on this diagram and the relationships between them. Spend a few moments now focusing on how I used the crosshair notation and the qualified name string notation in Figure 11.1 to specify where each of these elements is located in the model hierarchy. You will see that only two elements on this diagram—*Mission Requirements Specification* and *DellSat-77 System Requirements Specification*—are contained within the default namespace, the *Requirements* package. All the other elements shown are nested elsewhere in the model hierarchy.

11.4 Requirements

The notation for a requirement is a rectangle with the stereotype «requirement» preceding the name. A requirement has two properties—*id* and *text*—both of type *String*. Those strings are entirely user defined; SysML imposes no constraints on what they can be. (I refer you, though, to the *INCOSE Systems Engineering Handbook* for authoritative guidelines on how to write good text requirements.)

SysML does not dictate precisely what a requirement represents. It is legal for a single requirement in your model to represent an atomic requirement, a compound requirement, or an entire requirements specification. This decision is a matter of methodology. Your team should have a modeling standards and conventions document that captures this decision so that all members of the team are modeling consistently.

The requirements diagram in Figure 11.1 displays five requirements. Three of those—*Hohmann Transfer, Thruster Burn,* and *Altimetry*—represent atomic requirements. The other two—*Mission Requirements Specification* and *DellSat-77 System Requirements Specification*—represent entire requirements specifications (as their names imply).

If your team chooses to adopt the convention of using a requirement to represent an entire specification, I strongly recommend adopting the practice of naming the element appropriately to convey that information to your reader. A common alternative practice is to use a package to represent a requirements specification. In fact, most modeling tools do this by default when you import a requirements specification into your system model.

As mentioned, SysML requirements have only two properties: *id* and *text*. In daily practice, though, requirements commonly have other properties that you care about (such as *rationale, priority, verification method, criticality*). SysML does not define these properties for requirements, because they vary from organization to organization. You can, of course, define your own custom stereotype that specializes the predefined «requirement» stereotype and adds any additional properties your team may need. However, this is an advanced technique that goes beyond the scope of this book. *A Practical Guide to SysML: The Systems Modeling Language*, by Sanford Friedenthal, Alan Moore, and Rick Steiner, offers a comprehensive discussion of this topic.

Although SysML does not define a *rationale* property for requirements, it does define an alternative mechanism for documenting rationale. I provide a concrete example of this mechanism in Section 11.7, "Rationale."

11.5 Requirements Relationships

Capturing requirements in your system model is useful. However, the greater value lies in the relationships you create among the requirements and other model elements. There are six kinds of requirements relationships that you're likely to use in your modeling work: *containment*, *trace*, *derive requirement*, *refine*, *satisfy*, and *verify*.

Note

SysML defines a seventh requirement relationship—the *copy* relationship—but you'll rarely use it in daily practice.

These relationships establish requirements traceability within the system model, which is commonly a process requirement in systems engineering organizations. Practically speaking, however, capturing these relationships in your model will enable you to use your modeling tool to autogenerate requirement traceability and verification matrices (RTVMs) and perform automated downstream impact analysis when the requirements change (as they inevitably will). These capabilities are huge time-savers, and that directly translates into cost savings.

11.5.1 Containment Relationships

I discuss the concept of *containment* extensively in Chapter 10. I state that a package is a kind of namespace; it can contain other named elements within the model hierarchy. A requirement is also a kind of namespace; it, too, can contain other named elements within the model hierarchy. But it has a constraint that a package doesn't have: A requirement can contain only other requirements.

Chapter 10 also introduces the three notations you can use to convey namespace containment: crosshair notation, nesting notation, and qualified name string notation. In Section 11.3, "The Requirements Diagram Frame," I mention that you can use two of these notations—crosshair notation and qualified name string notation—to relate requirements. (Requirements do not have a graphical compartment, so the nesting notation is not an option.)

I use the crosshair notation in Figure 11.1 to convey that the *DellSat-77 System Requirements Specification* element contains the *Hohmann Transfer* requirement. I use the qualified name string notation to convey that the *Propulsion Subsystem Requirements Specification* element contains the *Thruster Burn* requirement. Similarly, a qualified name string conveys that the *Sensor Payload Requirements Specification* element contains the *Altimetry* requirement.

Those two qualified name strings do not begin with the name of the model. We therefore refer to them as *relative* qualified name strings; the path begins with the default namespace for this diagram: the *Requirements* package.

In contrast, the test case, use case, and block that are shown on this diagram display qualified name strings that *do* begin with the name of the model; we refer to those as *fully* qualified name strings. (I discuss these three elements in more detail in the context of the other requirements relationships shortly.)

11.5.2 Trace Relationships

Formally, the **trace** relationship is a kind of dependency. The notation for a trace relationship is the same as the one for a dependency—a dashed line with an open arrowhead—except that it has the «trace» stereotype applied to it.

I use the trace relationship in Figure 11.1 to convey that the *DellSat-77 System Requirements Specification* traces back to the *Mission Requirements Specification*. The trace relationship, however, is a weak relationship. It conveys nothing more than a basic dependency would: A modification to the supplier element (at the arrowhead end) *may* result in the need to modify the client element (at the tail end).

The trace relationship is still useful insofar as a modeling tool can navigate that relationship to autogenerate an RTVM or perform automated downstream impact analysis. In your daily practice, however, I recommend that you instead use one of the other (more meaningful) requirements relationships discussed shortly.

11.5.3 Derive Requirement Relationships

A **derive requirement** relationship is another kind of dependency. The notation for a derive requirement relationship is therefore the same as the one for a dependency, but it has the «deriveReqt» stereotype applied to it. This relationship must have a requirement at both the client and the supplier ends. The derive requirement relationship conveys that the requirement at the client end is derived from the requirement at the supplier end.

The requirements diagram in Figure 11.1 displays two derive requirement relationships. This model conveys that the *Thruster Burn* and *Altimetry* requirements are derived from the *Hohmann Transfer* requirement. *Thruster Burn* and *Altimetry* are the client elements in the relationships shown; *Hohmann Transfer* is the supplier.

It's entirely legal to have multiple levels of derivation. And dependencies are transitive. Therefore, if a primary requirement changes, the downstream impact may ripple through the entire chain of derive requirement relationships.

It's also important to note that a derived requirement does not need to be contained in the same namespace as its supplier; it may be nested elsewhere in the model hierarchy. That's the case in the model shown in Figure 11.1 (as conveyed by the containment relationship notations).

11.5.4 Refine Relationships

A **refine relationship** is another kind of dependency. The notation for a refine relationship is the same as the one for a dependency, but with the «refine» stereotype applied to it. A refine relationship conveys that the element at the client end is more concrete (i.e., less abstract) than the element at the supplier end.

SysML imposes no constraints on the kinds of elements that can appear at either end of a refine relationship. The common convention, however, is to refine a textual, functional requirement by using a use case. A use case—more precisely, its accompanying specification—provides detail and clarity that a text requirement by itself cannot express.

The requirements diagram in Figure 11.1 conveys that the *Measure altitude* use case refines the *Altimetry* requirement. The text use case specification (or the activity diagram) that accompanies the use case contains a detailed sequence of steps that specifies the required behavior of the system more precisely than the *Altimetry* requirement does by itself.

11.5.5 Satisfy Relationships

A **satisfy relationship** is another kind of dependency. The notation for a satisfy relationship is the same as the one for a dependency, but with the «satisfy» stereotype applied to it. This relationship must have a requirement at the supplier end. SysML imposes no constraints on the kind of element that can appear at the client end. By convention, however, the client element is always a block.

The requirements diagram in Figure 11.1 conveys that the *Microcosm Autonomous Navigation System (MANS)* block satisfies the *Altimetry* requirement. The satisfy relationship is an assertion that an instance of the block at the client end will fulfill the requirement at the supplier end.

An important point, however, is that an assertion doesn't constitute proof. The satisfy relationship is simply a mechanism to allocate a requirement to a structure. The proof of that satisfaction will come from test cases.

11.5.6 Verify Relationships

A **verify relationship** is another kind of dependency. The notation for a verify relationship is the same as the one for a dependency, but with the «verify» stereotype applied to it. Like the satisfy relationship, the verify relationship must have a requirement at the supplier end. SysML imposes no constraints on the kind of element that can appear at the client end. By convention, however, the client element is always a test case.

A **test case** is simply a behavior that you define somewhere in your model—one that you create for the purpose of invoking a particular structure's functionality to verify that it satisfies one or more of the requirements allocated to it. A test case can be any one of the three kinds of behaviors: activity, interaction, or state machine. However, a test case is most often modeled as an interaction (and displayed on a sequence diagram).

The requirements diagram in Figure 11.1 conveys that the test case named *Hohmann Transfer Simulation, Main Success Scenario* verifies the

Hohmann Transfer requirement. This test case is a behavior that, when executed, will prove that the system implementation actually satisfies the *Hohmann Transfer* requirement. (Of course, a system model typically contains many sets of test cases to fully verify all system and component requirements.)

A comprehensive discussion of modeling test cases is beyond the scope of this book. When you're ready to expand your modeling repertoire into the testing domain, I recommend the excellent book *Model-Driven Testing: Using the UML Testing Profile* by Paul Baker et al.

11.6 Notations for Requirements Relationships

SysML offers a variety of notations to display the requirements relationships on diagrams: direct notation, compartment notation, callout notation, matrices, and tables. Each notation has pros and cons that make it better suited than the others depending on your needs and constraints. These notation options, however, are available only for the dependency-based requirements relationships: trace, derive requirement, refine, satisfy, and verify. (The containment relationship is limited to the crosshair notation and qualified name string notation, which I discuss earlier.)

11.6.1 Direct Notation

Direct notation refers to the dependency notation itself: a dashed line with an open arrowhead, with a stereotype applied to it to convey the specific relationship. There are several examples of this notation in Figure 11.1.

The advantage of this notation is that it puts the relationship itself into focus on the diagram (because it takes up so much of the reader's visual field). And that, of course, is its disadvantage; it consumes a lot of real estate on the diagram.

11.6.2 Compartment Notation

Any element that can display compartments (e.g., blocks, requirements, test cases) can use **compartment notation** to display requirements relationships. Each relationship appears in a separate compartment. The compartment name specifies both the kind of relationship and the direction with respect to the element at the other end.

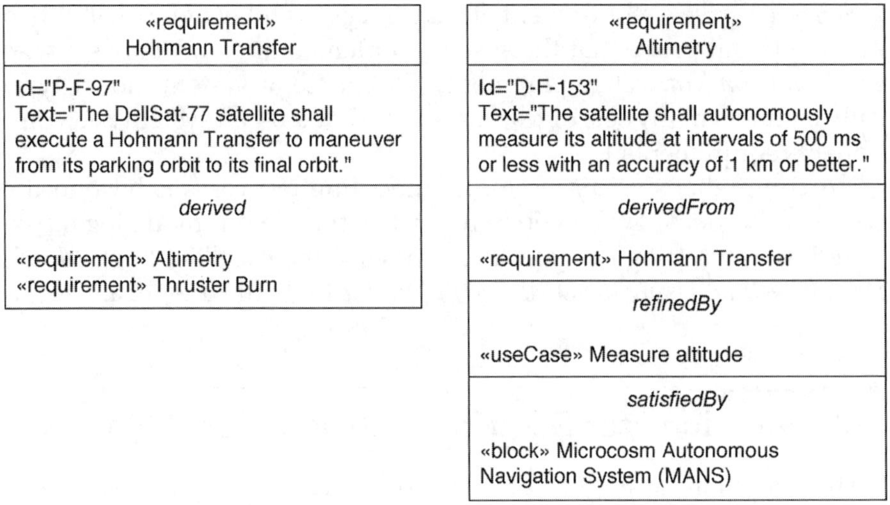

Figure 11.2 *Displaying requirements relationships using compartment notation*

The advantage of this notation is that it's more compact than direction notation and callout notation. Several requirements relationships can be displayed, all within the boundary of a single element (as shown in the *Altimetry* requirement in Figure 11.2). And each compartment can list multiple elements, as long as they all participate in the kind of relationship named in that compartment (as shown in the *Hohmann Transfer* requirement in Figure 11.2). This notation option is a good choice when you need to put the primary focus on the features of the elements, with a secondary focus on their relationships.

The disadvantage of this notation is that it's limited to the kinds of elements that can display compartments. And even though it's more compact than direct notation and callout notation, it still consumes more diagram real estate per relationship than do matrices and tables.

11.6.3 Callout Notation

Callout notation refers to a comment anchored to an element (see Figure 11.3). The contents of the callout notation are identical to the contents of the compartment notation: It specifies the kind of relationship and the direction with respect to the element at the other end. It then specifies the kind and name of that element at the other end. Like a single compartment, a single callout can list multiple elements if they all participate in that same kind of relationship.

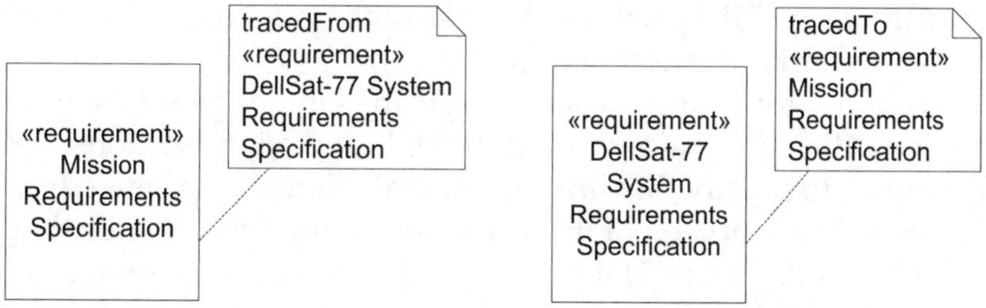

Figure 11.3 *Displaying requirements relationships using callout notation*

The advantage of this notation is its versatility; a callout can be anchored to any kind of element on any kind of diagram. The disadvantage is that it is the least space-efficient notation option.

11.6.4 Matrices

Matrices are a staple in systems engineering documentation. A **matrix** is not a graphical notation (as most SysML notations are). However, SysML supports matrix notation because it is simply the best mechanism for expressing many relationships in the least amount of space. The matrix shown in Figure 11.4 expresses all of the dependency-based requirements relationships that are displayed in Figure 11.1.

Now that you know that SysML supports matrix notation for requirements relationships, I caution you not to get attached to the particular format shown in Figure 11.4. Matrix format is not explicitly

Client	Supplier		
	«requirement» Mission Requirements Specification	«requirement» Hohmann Transfer	«requirement» Altimetry
«requirement» DellSat-77 System Requirements Specification	trace		
«requirement» Thruster Burn		deriveReqt	
«requirement» Altimetry		deriveReqt	
«testCase» Hohmann Transfer Simulation, Main Success Scenario		verify	
«useCase» Measure altitude			refine
«block» Microcosm Autonomous Navigation System (MANS)			satisfy

Figure 11.4 *Displaying requirements relationships using matrix notation*

defined in the SysML specification. Different modeling tools, therefore, implement matrix notation in different ways.

Some tools, for example, name a specific kind of relationship in the upper-left cell and show arrows in the cells at the intersections of rows and columns to specify the direction of that relationship between a pair of elements. This implementation is considerably less versatile than the format shown in Figure 11.4 because a given matrix is limited to displaying a single kind of relationship.

Matrix notation has two disadvantages to keep in mind when you make your decision. It does not display the features of the elements—only the relationships between them. And even though it's very space efficient and readable in your modeling tool (where the reader has scroll bars), it becomes much less readable when the need arises to insert it into printed documentation.

11.6.5 Tables

Tables, like matrices, are commonplace in systems engineering documentation. And like matrices, SysML supports tables because they are a space-efficient mechanism for expressing a lot of information. Tables are not quite as compact as matrices. However, table notation allows you to display the properties of elements (e.g., *id*) as well as the relationships between them, as shown in Figure 11.5.

Another similarity between tables and matrices is that no particular table format is defined in the SysML specification. The implementation

id	name	relation	id	type	name
SRS-1081	DellSat-77 System Requirements Specification	trace	MRS-1016	requirement	Mission Requirements Specification
P-F-97	Hohmann Transfer	verifiedBy		testCase	Hohmann Transfer Simulation, Main Success Scenario
D-F-152	Thruster Burn	deriveReqt	P-F-97	requirement	Hohmann Transfer
D-F-153	Altimetry	deriveReqt	P-F-97	requirement	Hohmann Transfer
		refinedBy		useCase	Measure altitude
		satisfiedBy		block	Microcosm Autonomous Navigation System (MANS)

Figure 11.5 *Displaying requirements relationships using table notation*

of the table notation is left to the discretion of the various modeling tool vendors.

Commercial-grade modeling tools let you define your own custom table format, including the insertion of extra columns for any properties that you need to display. Keep in mind, however, that a table becomes cumbersome to read in printed documentation when its width spans multiple pages.

11.7 Rationale

As mentioned earlier in Section 11.4, "Requirements," a SysML requirement has two properties: *id* and *text*. The language, unfortunately, does not define a "rationale" property for requirements. It does, however, provide an alternative mechanism to document rationale in your model—a mechanism that applies to all kinds of elements and relationships, and not only requirements.

Rationale is captured in a SysML model as a specialized kind of comment. Recall from Section 3.11, "Comments," that the notation for a comment is a note symbol—a rectangle with the upper-right corner bent. The notation for a rationale element is simply a note symbol with the stereotype «rationale» preceding the body of the comment, as shown in Figure 11.6.

It is legal to attach a rationale to any kind of element and any kind of relationship between two elements. However, you're most likely to use this feature of the language in the context of requirements and allocations (I talk about the latter in detail in the next chapter).

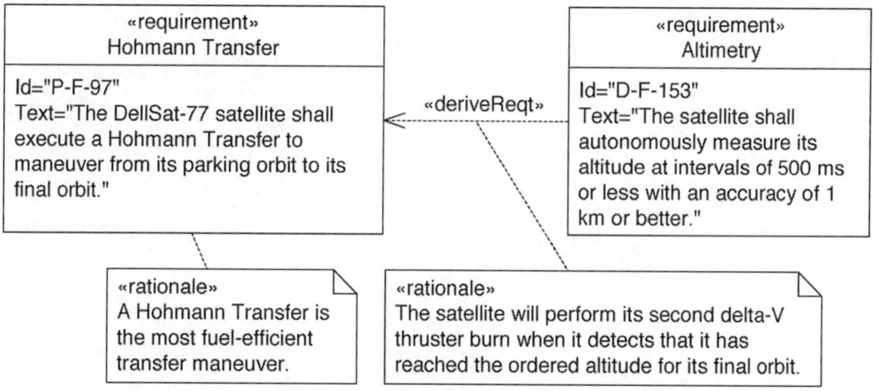

Figure 11.6 *Capturing rationale in a SysML model*

Summary

A requirements diagram is the primary medium in SysML for communicating system requirements that are specified in text. Modelers commonly create requirements diagrams to convey traceability among requirements and traceability from requirements to system structures and behaviors.

Capturing this traceability within a SysML model enables you to use a modeling tool to perform automated downstream impact analysis—generating a targeted list of system structures and behaviors that you may need to modify when the requirements they depend on change over time. This capability significantly reduces the time and cost to implement changes to a design over the course of the system life cycle.

Chapter 12

Allocations: Cross-Cutting Relationships

A useful relationship in system design is the allocation relationship. This relationship is not specific to any one of the nine kinds of SysML diagrams. It's a versatile relationship that you and your team can use in various ways across all aspects of a system model.

I begin with a discussion of the conventional ways in which systems engineers have used allocations historically. Then I present several notations that you can use on diagrams to express allocations among model elements, and I discuss the strengths and weaknesses of each option.

12.1 Purpose

Systems engineers often speak of **allocations** in the course of performing system design. The *INCOSE Systems Engineering Handbook* discusses allocations in the following contexts: allocating requirements to structures, allocating behaviors to structures, allocating logical structures to physical structures, and allocating resources to structures (e.g., mass, power, cost, throughput).

SysML supports the concept of allocations, enabling you to capture the results of allocation activities directly in your system model. Creating allocation relationships in your model ensures consistency in the flow-down of design inputs from one architectural level to the next and, more importantly, allows you to revisit those decisions efficiently and systematically as the life cycle progresses.

12.2 There's No Such Thing as an Allocation Diagram

Allocations are commonly referred to as cross-cutting relationships. You can create an allocation relationship between *any* two model elements, no matter where they are contained within your system model or which kinds of diagrams they appear on. An allocation relationship cuts across the various aspects of your model: requirements, behaviors, structures, and constraints. To facilitate this, SysML allows you to display allocation relationships on any of the nine kinds of SysML diagrams. There's no diagram kind devoted specifically to this relationship.

12.3 Uses for Allocation Relationships

The SysML specification does not define what any particular usage of an allocation relationship actually means; it's intentionally silent on this point. The specification simply states the following:

> The allocation relationship can provide an effective means for navigating the model by establishing cross relationships, and ensuring the various parts of the model are properly integrated. [SysML] does not try to limit the use of the term "allocation," but provides a basic capability to support allocation in the broadest sense.

This flexibility in the language gives you the freedom to use allocations in any way that is meaningful for your project team. The key word there, of course, is *meaningful*.

The modeling methodology that your team adopts will specify meaningful uses for allocations. These uses should be explicitly captured in your team's modeling standards and conventions document to ensure that all members of the team are using allocations consistently in the system model.

In the remainder of this section, I discuss some common uses for allocations and show concrete examples of this relationship. Keep in mind, however, that the language does not specify any meaning for these uses. These are simply conventional ways that systems engineers use allocations.

12.3.1 Behavioral Allocation

Behavioral allocation (also known as **functional allocation**) refers generally to the activity of allocating a behavioral element to a structural element. Most often, this takes one of two specific forms:

- Allocating an activity, an interaction, or a state machine behavior to a block
- Allocating an action (in an activity) to a part property (owned by a block)

The notation for an allocation relationship is the same as the one for a dependency—a dashed line with an open arrowhead—but with the «allocate» stereotype applied to it. The element being allocated appears at the tail end of the line; the element receiving the allocation appears at the arrowhead end of the line.

The BDD in Figure 12.1 conveys that the *Measure altitude* activity and the *Determine attitude* activity are allocated to the *Microcosm Autonomous Navigation System (MANS)* block. These relationships tell the reader that all instances of this block will be able to perform these two behaviors (for any client that requires them).

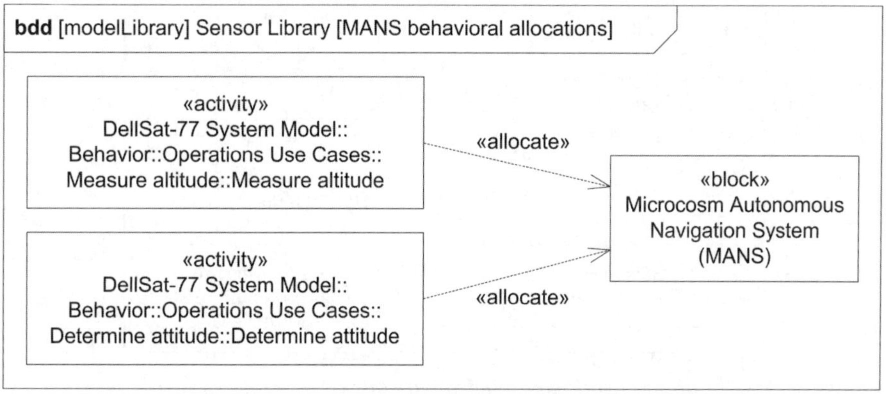

Figure 12.1 *Allocations of activities to a block (behavioral allocation)*

I provide a concrete example of the other form of behavioral allocation—actions to part properties—when I discuss allocation activity partitions in Section 12.4.6, "Allocation Activity Partitions."

12.3.2 Structural Allocation

Structural allocation refers to the activity of allocating a structural element to another structural element. Common forms of structural allocation include the following:

- Allocating a logical block to a physical block
- Allocating a software property to a hardware property

It's a common technique in system design to define a logical system architecture early in the life cycle to better understand *what* the system must do and then later define one or more candidate physical architectures that demonstrate *how* the system will do it. If your team has adopted this methodology, you can use the SysML allocation relationship to map elements in the logical architecture to their corresponding elements in the physical architecture.

Specifying the deployment of software components to hardware nodes is a common need in system development. UML offers a kind of diagram—the deployment diagram—dedicated to this purpose; SysML does not. Instead, you can use the allocation relationship to meet this need in a SysML model.

The IBD in Figure 12.2 displays two allocation relationships from software components to hardware nodes. Allocation relationships are

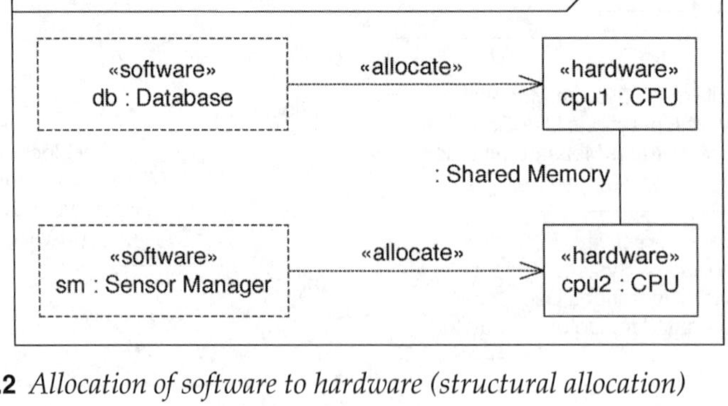

Figure 12.2 *Allocation of software to hardware (structural allocation)*

used in this context to convey the deployment of software in a distributed system.

The «software» and «hardware» stereotypes are not defined in SysML; they are custom stereotypes that modeling teams commonly define and apply to blocks, part properties, and reference properties to clarify the nature of those elements. The allocations shown in Figure 12.2 are from reference properties to part properties—all of which are structural elements. These allocations, therefore, are examples of structural allocations.

12.3.3 Requirements Allocation

Requirements allocation—the activity of allocating requirements to structures—is perhaps the most common form of allocation. Ironically, you don't use the SysML allocation relationship to perform requirements allocation in a system model; instead, you use one of the requirements relationships discussed in Chapter 11, "Requirements Diagrams"—specifically, the «satisfy» relationship.

When you create a «satisfy» dependency from a block to a requirement, you're asserting that all instances of that block will, in fact, satisfy that requirement. You are, in effect, allocating that requirement to that block. Take a look at Figure 11.1 to see a concrete example of this relationship.

12.4 Notations for Allocation Relationships

All the notation options discussed in Chapter 11 are also available for allocations: direct notation, compartment notation, callout notation, matrices, and tables. And there's one additional notation option that you can use for allocations that you can't use for the requirements relationships: allocation activity partitions.

12.4.1 Direct Notation

Direct notation refers to the notation shown in Figures 12.1 and 12.2. This notation puts a strong focus on the relationship itself. However, its use is limited to allocations between pairs of elements that you can legally display on the same diagram. Given the nature of an allocation as a cross-cutting relationship, that isn't always the case.

12.4.2 Compartment Notation

You can use **compartment notation** with any kind of element that can display compartments (e.g., blocks and activities on BDDs; part properties and reference properties on IBDs). The compartment name tells the reader the direction of the allocation relationship. If the compartment name is *allocatedTo*, then the elements listed in the compartment are at the arrowhead end of the relationship (i.e., they are the targets of the allocation). If the compartment name is *allocatedFrom*, then the elements listed in the compartment are at the tail end of the relationship (i.e., they are the elements being allocated). Figure 12.3 provides an example of compartment notation.

Compartment notation is more compact than direct notation or callout notation. You can list multiple elements in a single compartment (as long as they're all at the same end of an allocation relationship with respect to the element that owns the compartment).

The drawback to compartment notation is that its use is limited to the kinds of elements that can display compartments (e.g., blocks, part properties, activities). This notation is not an option for the kinds of elements that can't display compartments (e.g., actors, item flows).

12.4.3 Callout Notation

As discussed in Chapter 11, **callout notation** is simply a comment (i.e., a note symbol) anchored to an element (see Figure 12.4). The contents of the callout notation are identical to the contents of the compartment notation: It specifies the direction of the allocation relationship and lists the elements at the other end.

Callout notation is the most versatile notation option; a callout can be anchored to any kind of element on any kind of diagram. However, it's the least space-efficient option.

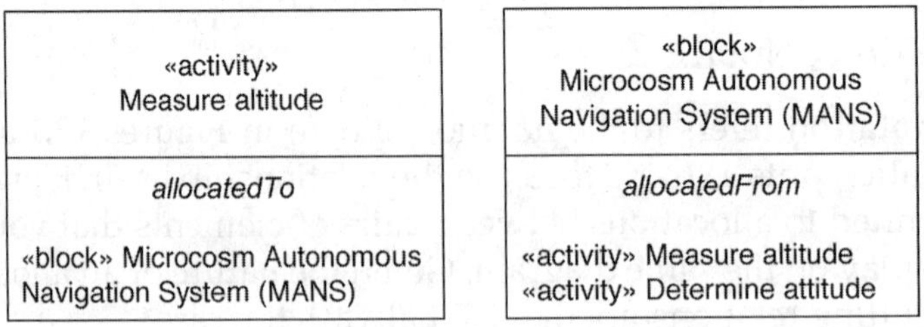

Figure 12.3 *Displaying allocation relationships using compartment notation*

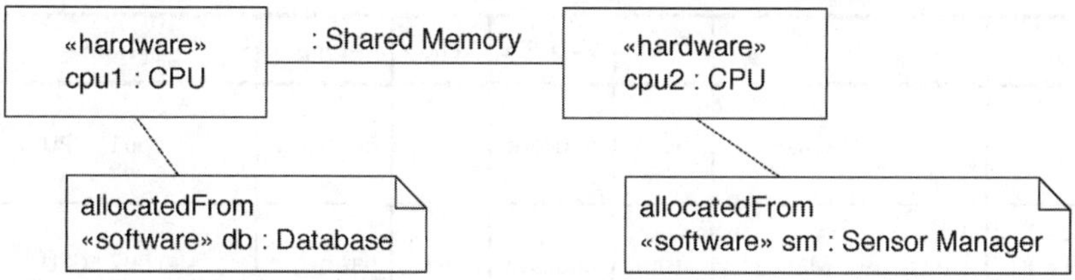

Figure 12.4 *Displaying allocation relationships using callout notation*

12.4.4 Matrices

A matrix is a traditional medium for expressing relationships in a system design. SysML supports matrix notation because it's the most efficient way to express a large number of relationships in the least amount of space.

I mention in Chapter 11 that the SysML specification does not explicitly define a format for matrix notation. Tool vendors are free to implement matrix notation in their own way in their respective modeling tools. The format shown in Figure 12.5 will likely differ from what you see in your particular modeling tool.

Matrix notation is very compact, but you should be aware of its limitations. It can display the relationships between elements but not the features of each element. It's also difficult to preserve the readability of a large matrix when you insert it into printed documentation.

12.4.5 Tables

A **table** is another traditional medium for expressing relationships in a system design. Tables are not quite as compact as matrices. However,

Source	Target		
	«block» Flight Computer	«block» Propulsion Subsystem	«block» Microcosm Autonomous Navigation System (MANS)
«activity» Measure altitude			allocate
«activity» Determine attitude			allocate
«interaction» Fire thrusters		allocate	
«activity» Validate command	allocate		

Figure 12.5 *Displaying allocation relationships using matrix notation*

Type	Name	End	Relation	End	Type	Name
software	db : Database	from	allocate	to	hardware	cpu1 : CPU
software	sm : Sensor Manager	from	allocate	to	hardware	cpu2 : CPU
software	vt : Vector Table	from	allocate	to	hardware	ic : Interrupt Controller

Figure 12.6 *Displaying allocation relationships using table notation*

table notation lets you add multiple columns to display the features of elements as well as the relationships between them (see Figure 12.6).

As with matrix notation, the SysML specification is silent on a format for table notation. The specific implementation of table notation is again left to the discretion of tool vendors. Modeling tools typically allow you to define your own custom table format, enabling you to add as many columns as you need. However, a wide table will be difficult to read when you insert it into printed documentation.

12.4.6 Allocation Activity Partitions

I discuss activity partitions in Chapter 6, "Activity Diagrams." I state that activity partitions are mechanisms for allocating behaviors to structures. An **allocation activity partition** is an activity partition with the «allocate» stereotype preceding the name in the header. The distinction between an activity partition with the «allocate» stereotype and one without the stereotype is subtle and has no impact on your daily modeling work; it would be perfectly fine if you and your team regarded all activity partitions as allocation activity partitions, whether or not you've applied the stereotype.

An allocation activity partition is a space-efficient notation for conveying allocation relationships; you can express many allocations on a single diagram. Its drawback, however, is its limited applicability. You can use an allocation activity partition only on an activity diagram, and you can use it only to perform one kind of allocation: behavioral allocation (an allocation of a behavioral element to a structural element).

With an allocation activity partition, the behavioral element that's getting allocated is always a node in an activity (most commonly, an action); the structural element that's receiving the allocation—which is named in the partition header—may be either a block or a part property. If the structural element is a block, then the allocation applies to *all* instances of that block. If the structural element is a part property, then the allocation applies only to the set of instances that the part property represents within a built system. (Recall from Chapter 3, "Block Definition Diagrams," that the number of instances a part property can represent is specified by its multiplicity.)

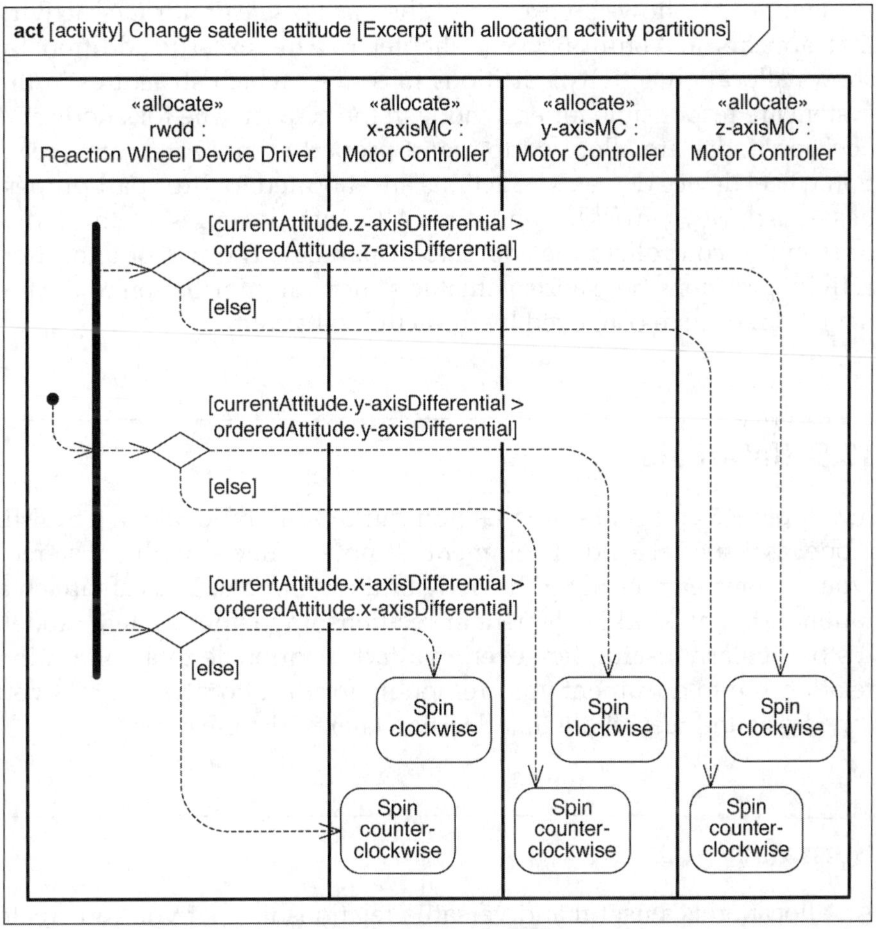

Figure 12.7 *Displaying allocation relationships using allocation activity partitions*

Note

Although the language doesn't require it, I recommend that you adopt the practice of having all partitions in a given activity represent either part properties or blocks. Don't mix the two.

Despite its limited applicability, an allocation activity partition is a common and useful way to express behavioral allocations. It enhances the narrative power of activity diagrams; stakeholders often value knowing who or what performs each action and not only which actions are performed.

Figure 12.7 shows an excerpt of the *Change satellite attitude* activity that appears in Figure 6.25 in Chapter 6. This excerpt additionally shows allocation activity partitions to convey which structures in the system are responsible for each node in the activity. The fork node and decision nodes are allocated to a part property that represents a reaction wheel device driver. The actions are allocated to three part properties—*x-axisMC*, *y-axisMC*, and *z-axisMC*—which represent three distinct motor controllers that operate in parallel. The use of allocation activity partitions here adds valuable structural information to the behavioral narrative conveyed by the activity diagram.

12.5 Rationale

In Chapters 3 and 11, I state that you can capture rationale in a SysML model as a special kind of comment—a note symbol—with the stereotype «rationale» preceding the body of the comment. You can attach a rationale to any kind of element or relationship in your system model. It's particularly useful, however, to attach a rationale to an allocation relationship to document your reasoning for the allocation so that you can revisit and reevaluate that decision later in the life cycle.

Summary

An allocation is a useful and versatile relationship that you can create between any two model elements—no matter where they exist in a system model or which diagrams they appear on. For this reason, we refer to allocations as cross-cutting relationships.

Systems engineers have traditionally used allocations for three specific purposes in system design—behavioral allocation, structural allocation, and requirements allocation—as defined in the *INCOSE Systems Engineering Handbook*. SysML itself, however, imposes no constraints on how you can use this relationship in a system model; you are free to define new uses for this relationship in any way that is meaningful for you and your stakeholders.

Appendix A

SysML Notation Desk Reference

This appendix provides a concise summary of the SysML graphical notations covered in this book. The notations are presented in tables, organized by diagram kind. The order of the tables corresponds to the order of the associated chapters in this book.

Each table has three columns: "Element Kind," "Notation," and "Section." The "Element Kind" column provides the formal name for the kind of element shown in that row. The "Notation" column shows the graphical representation of that kind of element. The "Section" column references the section number in the book where I discuss that kind of element in detail.

Table A.1 *Graphical Notations for Block Definition Diagrams (BDDs)*

Element Kind	Notation	Section
Block	«block» Block Name *parts* partPropertyName : Type [multiplicity] *references* referencePropertyName : Type [multiplicity] *values* valuePropertyName : Type [multiplicity] = default value *constraints* constraintPropertyName : Type *operations* operationName(parameterName : Type, ...) : returnType [multiplicity] *receptions* «signal» receptionName(parameterName : Type, ...)	3.4
Actor	 Actor Name A «actor» Actor Name B	3.8
Flow Specification	«flowSpecification» Flow Specification Name *flowProperties* direction flowPropertyName : Type	3.4.1.5
Constraint Block	«constraint» Constraint Block Name *constraints* {constraint expression} *parameters* parameterName : Type [multiplicity]	3.4.1.4, 3.10

Table A.1 *(continued)*

Element Kind	Notation	Section
Interface	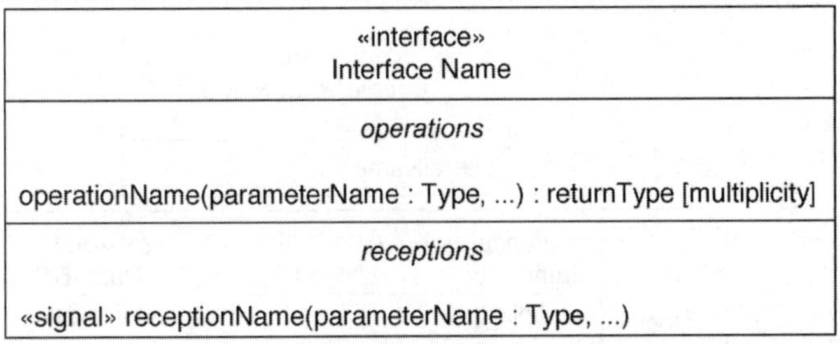	3.4.1.5
Signal	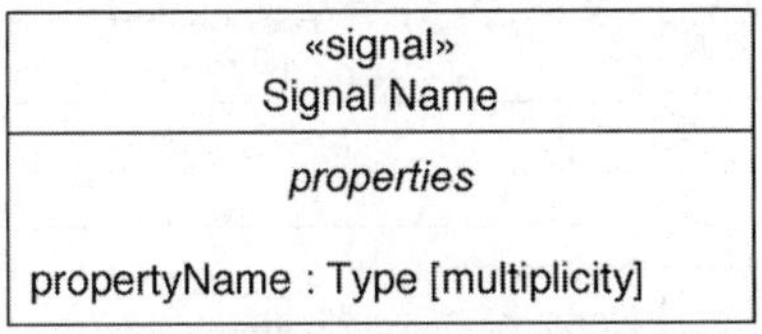	3.4.2.2
Standard Port (with Provided and Required Interfaces)	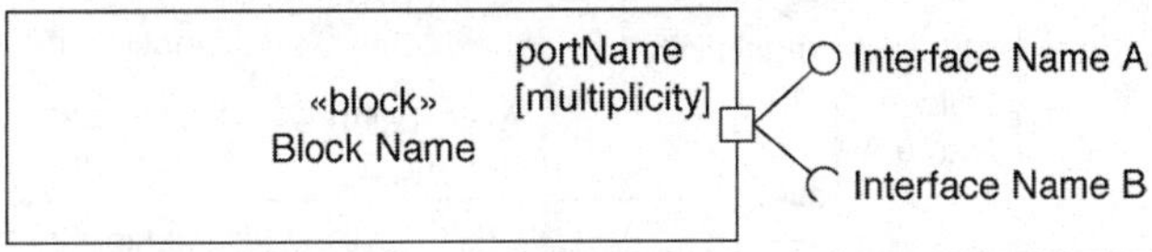	3.4.1.5
Atomic Flow Port	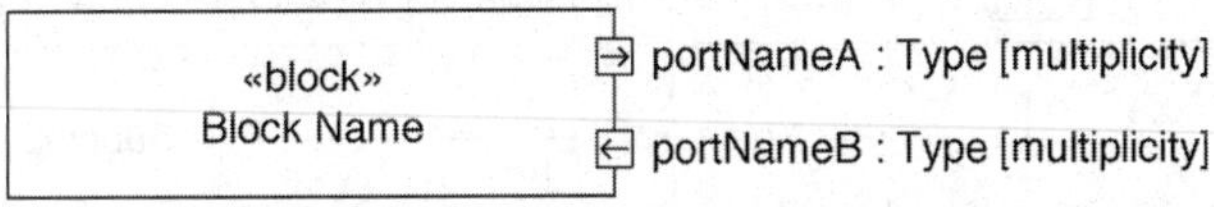	3.4.1.5
Nonatomic Flow Port	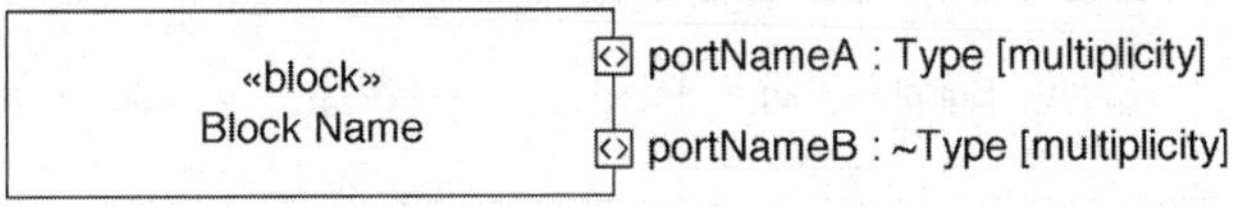	3.4.1.5
Comment	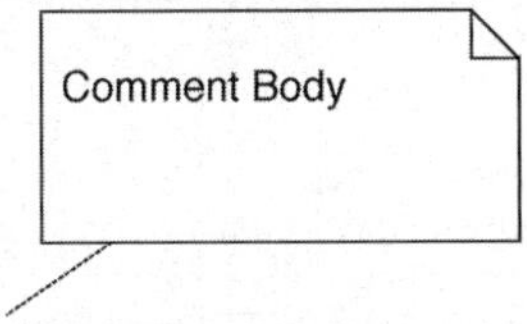	3.11
Value Type	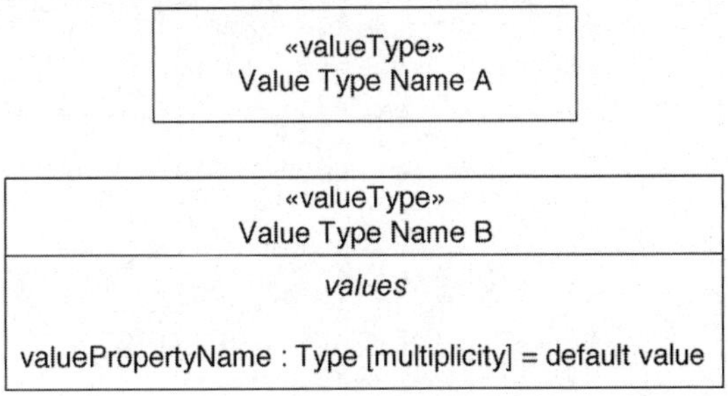	3.4.1.3, 3.9

(continues)

Table A.1 *(continued)*

Element Kind	**Notation**	**Section**
Enumeration	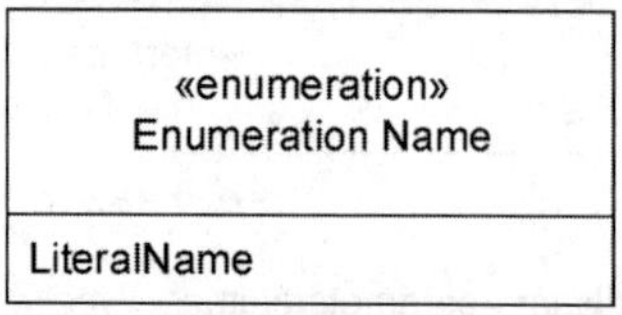	3.9
Reference Association	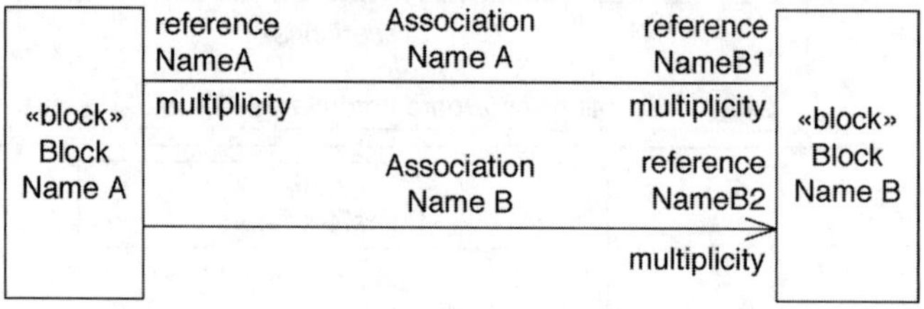	3.5.1
Composite Association	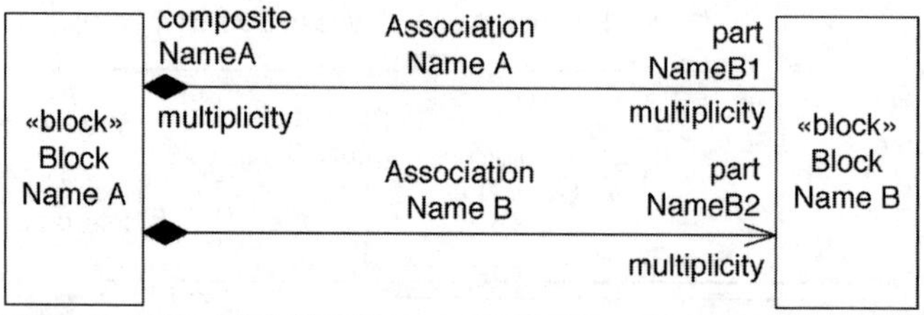	3.5.2
Generalization	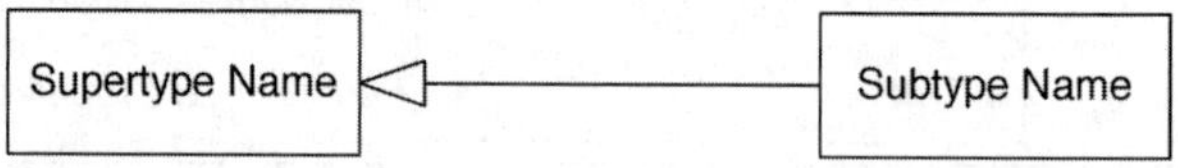	3.6
Dependency		3.7

Table A.2 *Graphical Notations for Internal Block Diagrams (IBDs)*

Element Kind	**Notation**	**Section**
Part Property	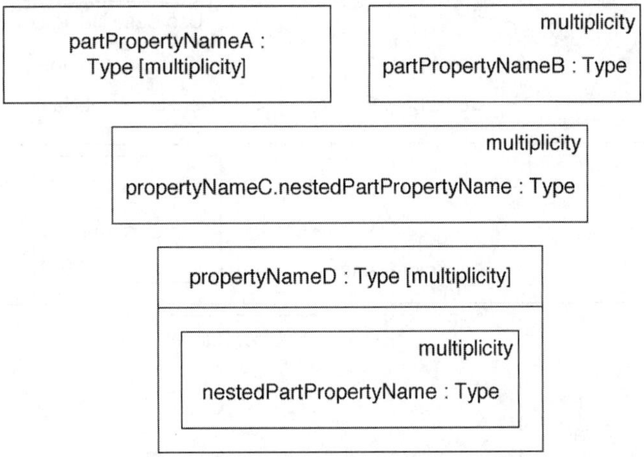	4.6, 4.10
Reference Property	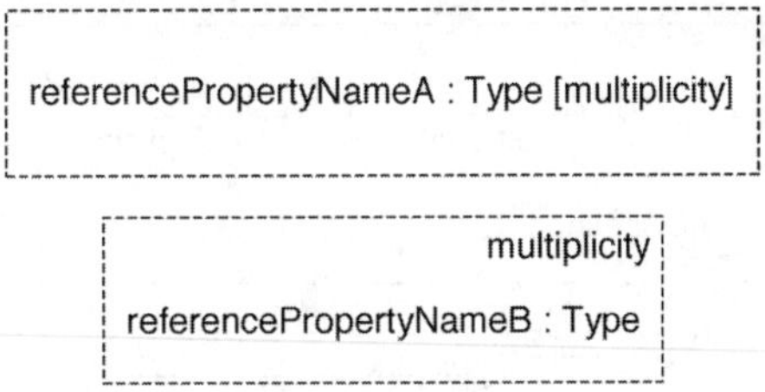	4.7, 4.10
Connector	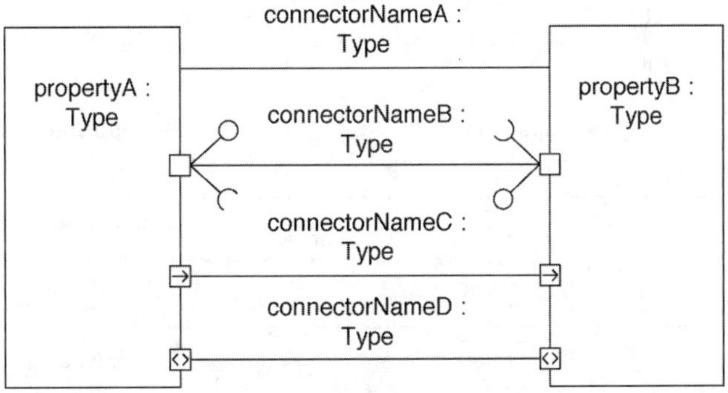	4.8, 4.9, 4.10
Item Flow	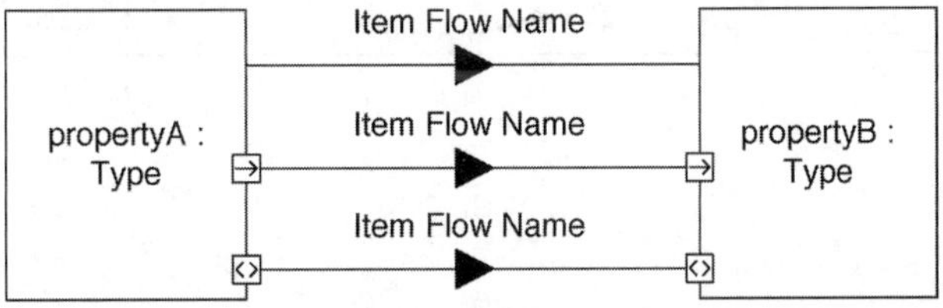	4.9

Table A.3 *Graphical Notations for Use Case Diagrams*

Element Kind	Notation	Section
Use Case	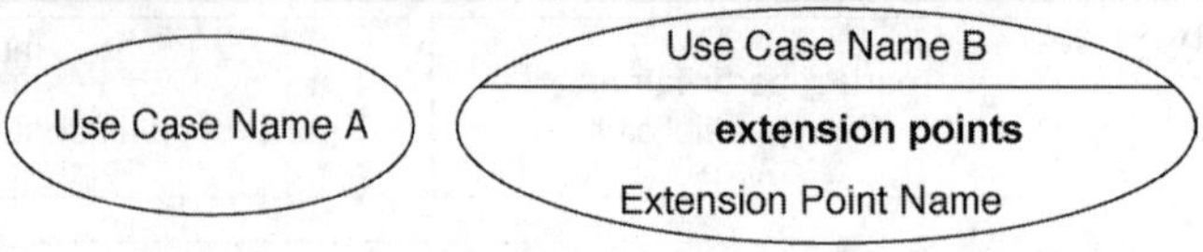	5.5, 5.8
Actor	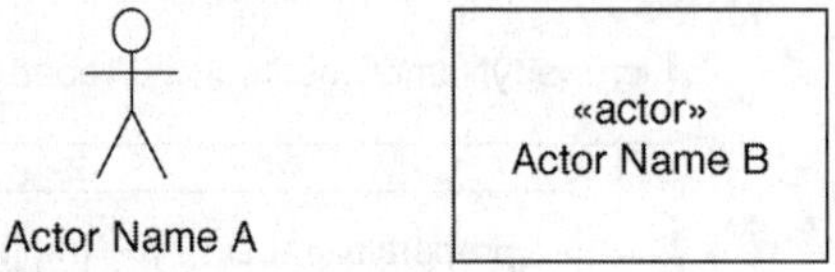	5.7, 5.8
System Boundary (or Subject)	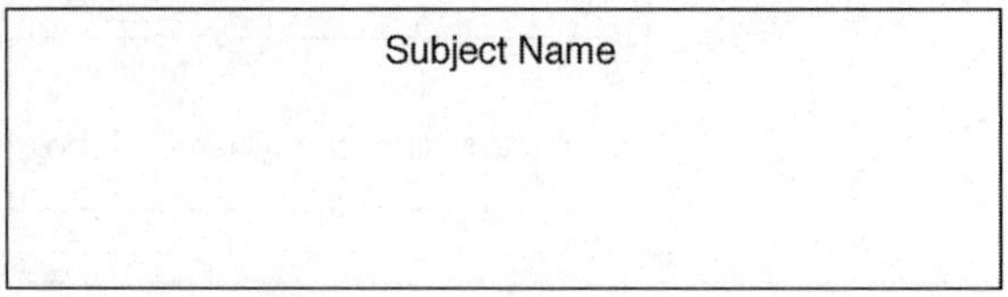	5.6
Reference Association	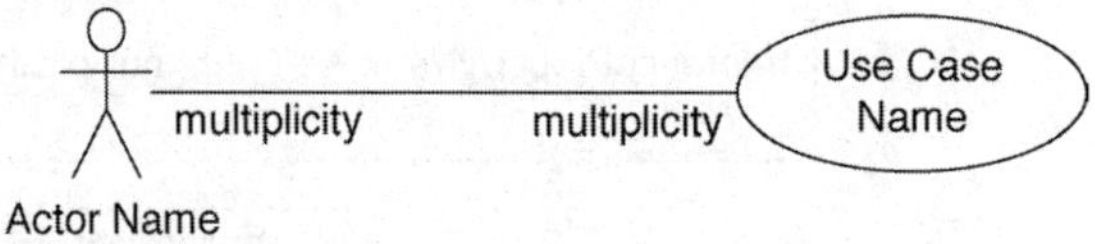	5.8
Generalization	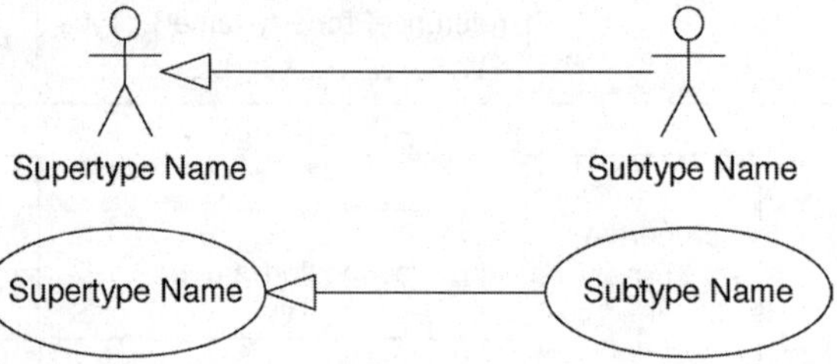	5.5, 5.7
Include		5.10
Extend	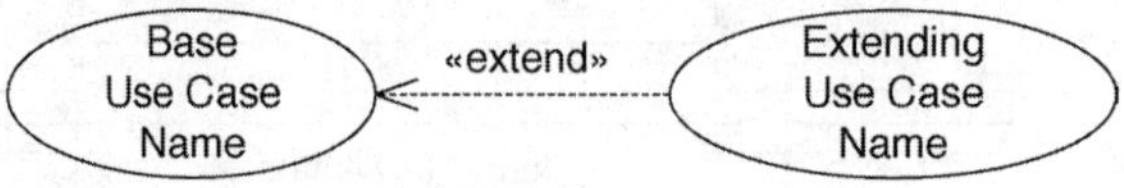	5.11

Table A.4 *Graphical Notations for Activity Diagrams*

Element Kind	**Notation**	**Section**
Basic Action		6.5
Object Node	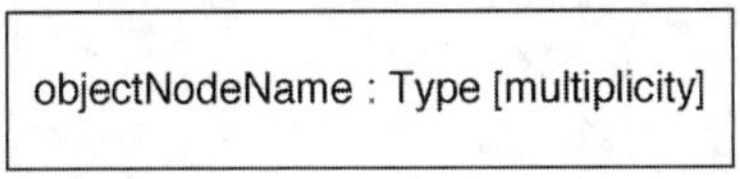	6.6, 6.6.3
Pin	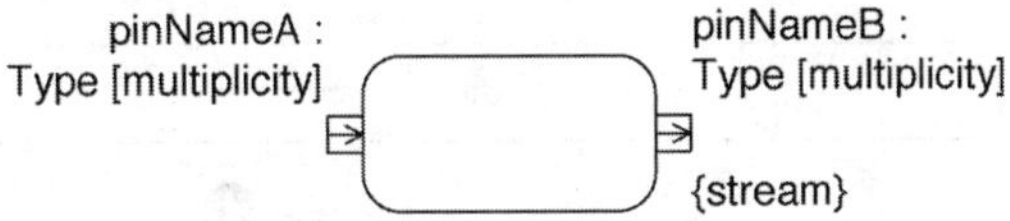	6.6.1, 6.6.3
Activity Parameter	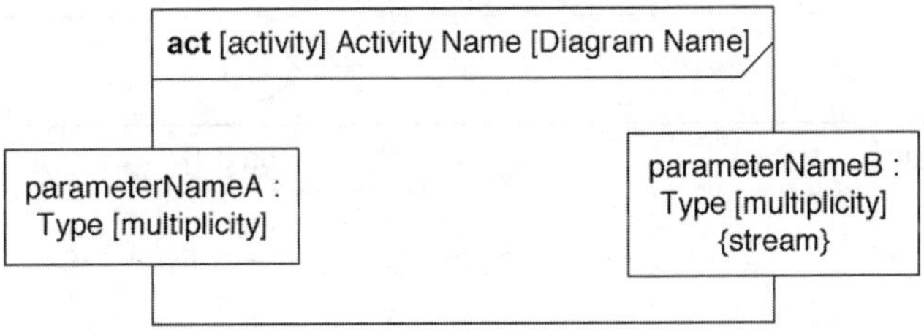	6.6.2, 6.6.3
Object Flow	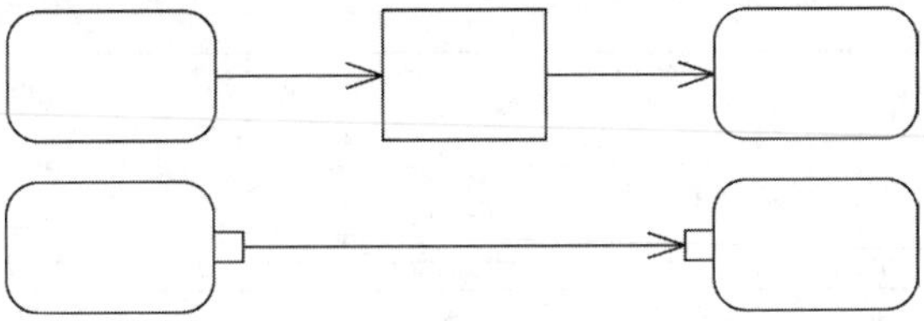	6.7.1
Control Flow	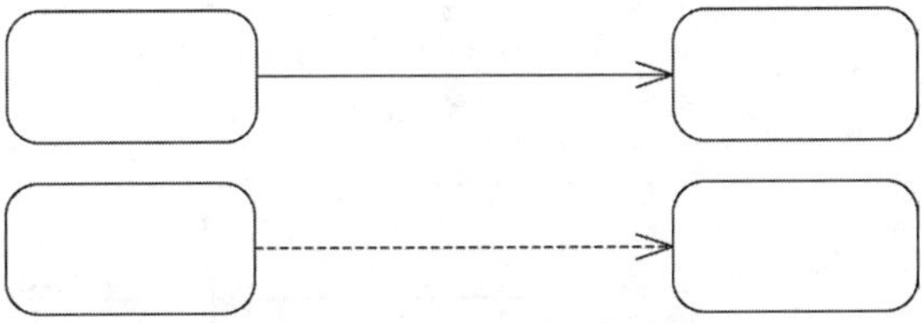	6.7.2
Call Behavior Action	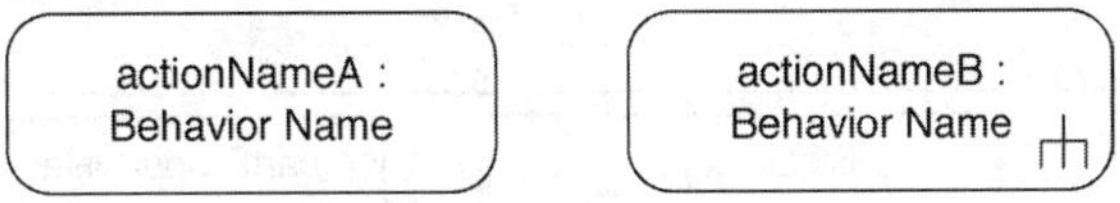	6.8.2
Send Signal Action		6.8.3

(continues)

Table A.4 *(continued)*

Element Kind	Notation	Section
Accept Event Action		6.8.4
Wait Time Action	 at (Absolute Time Expression) after (Relative Time Expression)	6.8.4.1
Initial Node		6.9.1
Activity Final Node		6.9.2
Flow Final Node	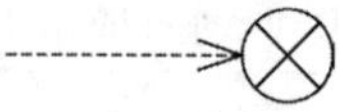	6.9.2
Decision Node	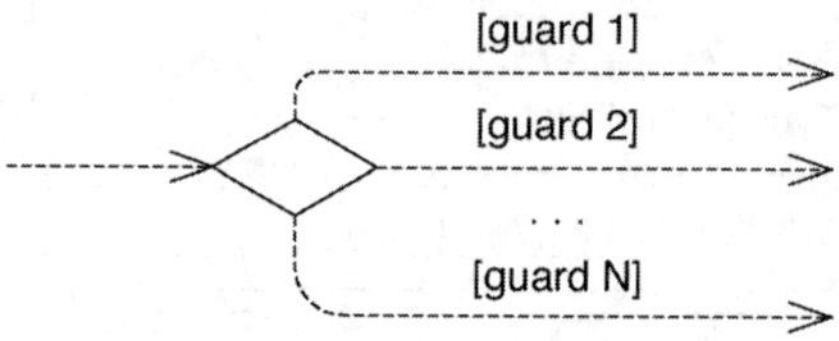	6.9.3
Merge Node	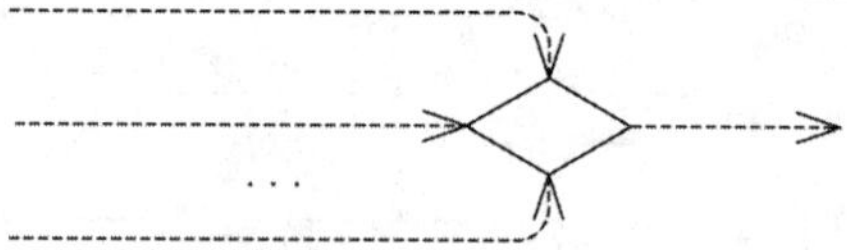	6.9.4
Fork Node	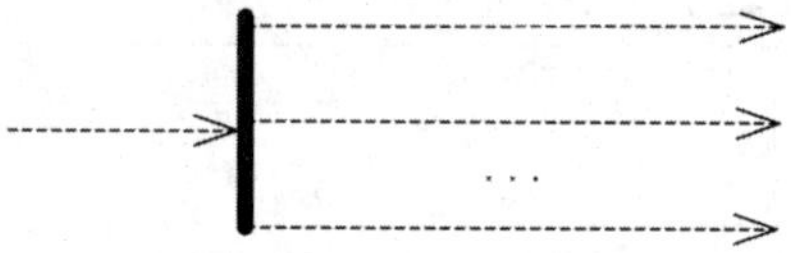	6.9.5
Join Node	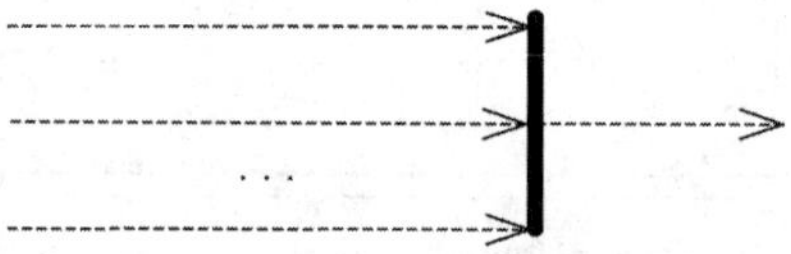	6.9.6
Activity Partition	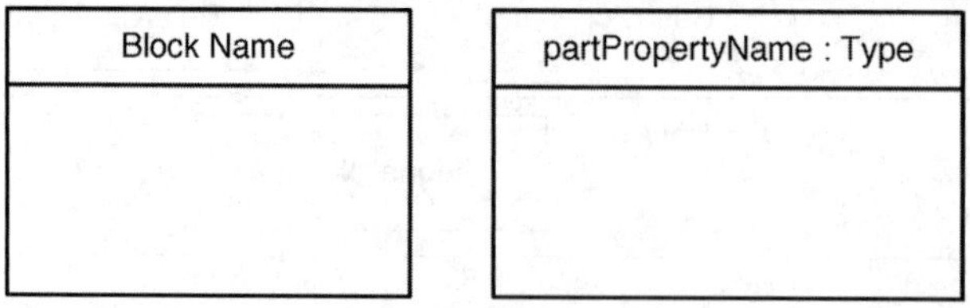	6.10

Table A.5 *Graphical Notations for Sequence Diagrams*

Element Kind	Notation	Section
Lifeline	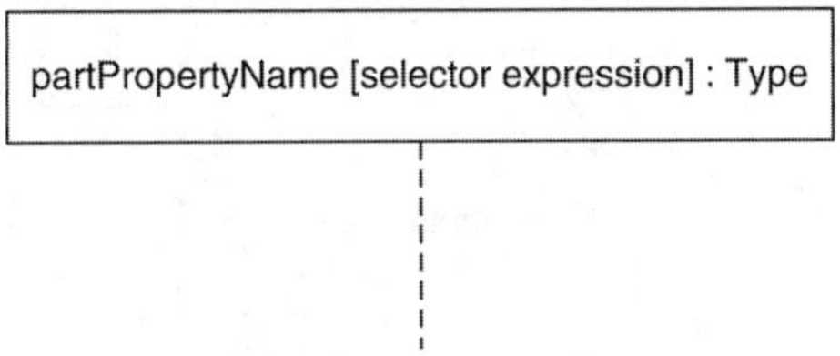	7.4
Asynchronous Message	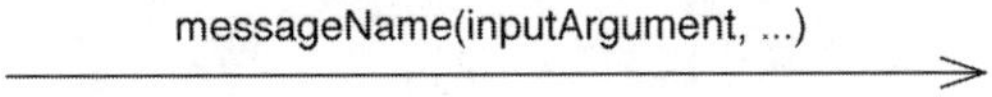	7.5.2.1
Synchronous Message	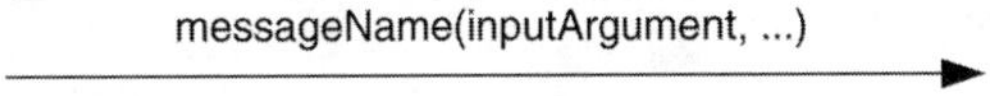	7.5.2.1
Reply Message	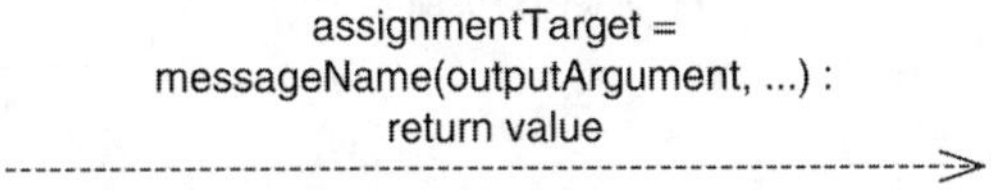	7.5.2.3
Create Message	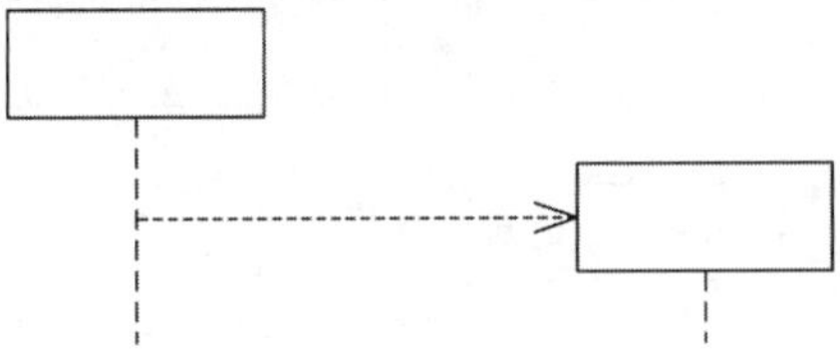	7.5.2.4
Destruction Occurence	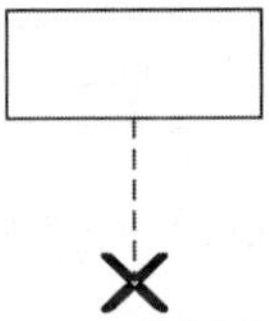	7.6
Execution Specification	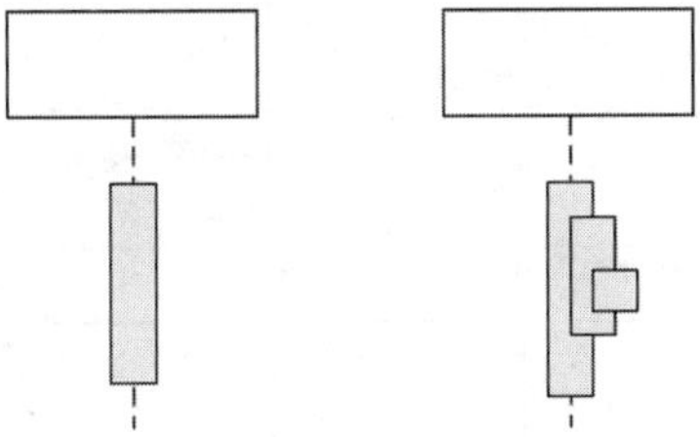	7.7
Time Constraint	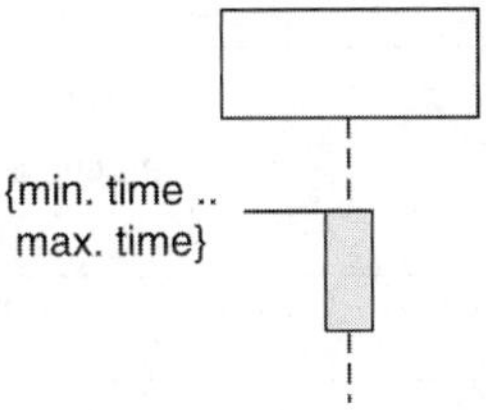	7.8.1

(continues)

Table A.5 *(continued)*

Element Kind	**Notation**	**Section**
Duration Constraint	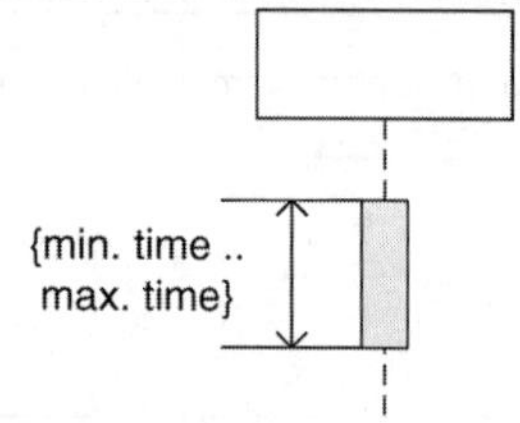	7.8.2
State Invariant	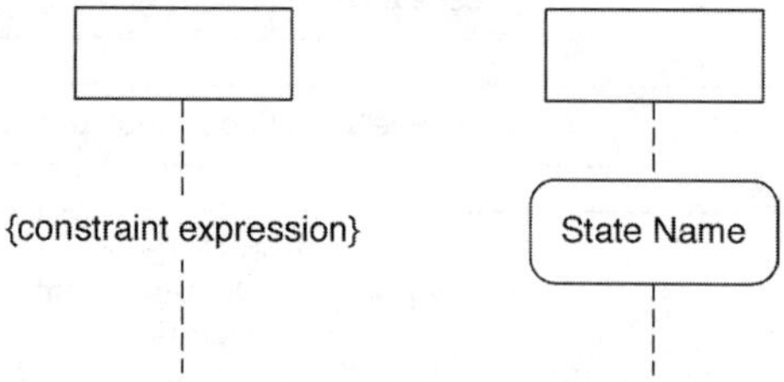	7.8.3
Combined Fragment	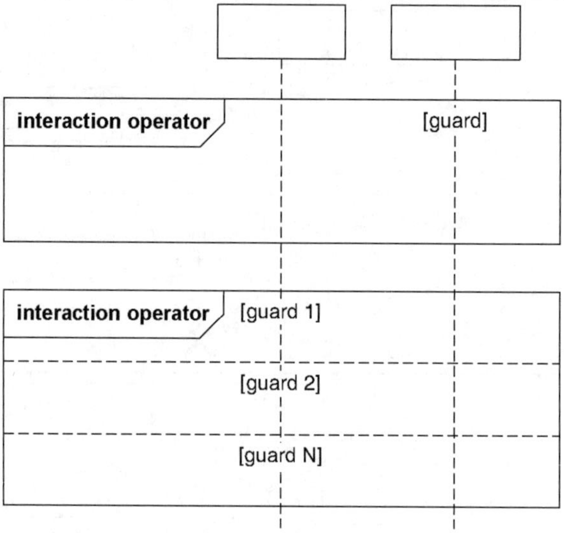	7.9
Interaction Use	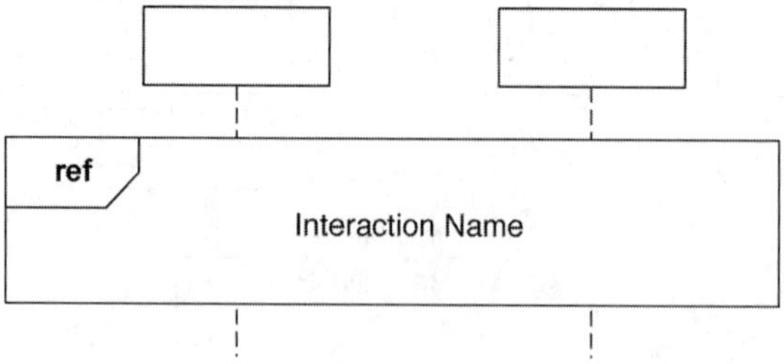	7.10

Table A.6 *Graphical Notations for State Machine Diagrams*

Element Kind	Notation	Section
Simple State	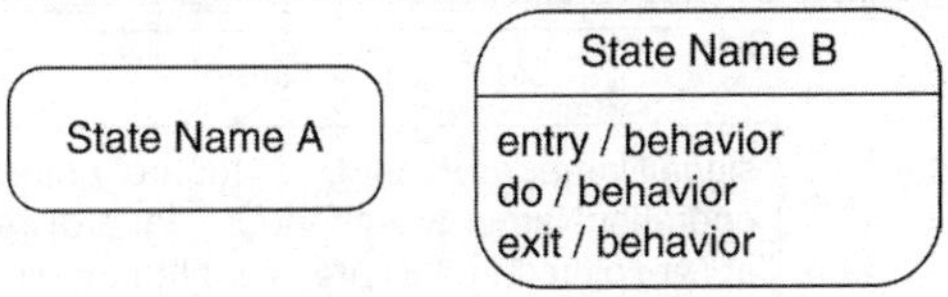	8.4.1
Composite State	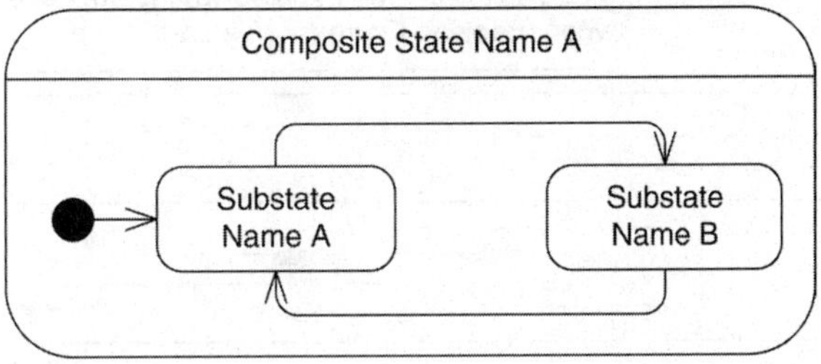 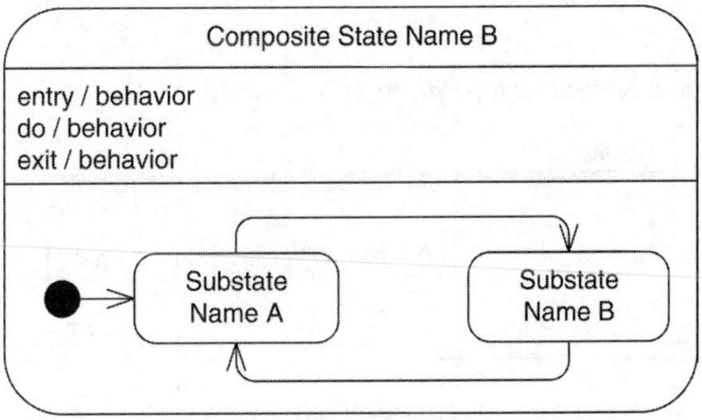	8.4.2
Final State	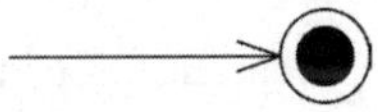	8.4.3
External Transition	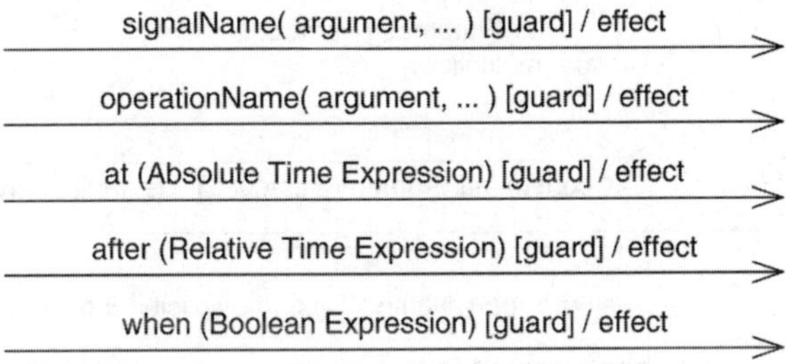	8.5, 8.5.2.1, 8.5.2.2, 8.5.2.3, 8.5.2.4

(continues)

Table A.6 *(continued)*

Element Kind	**Notation**	**Section**
Internal Transition	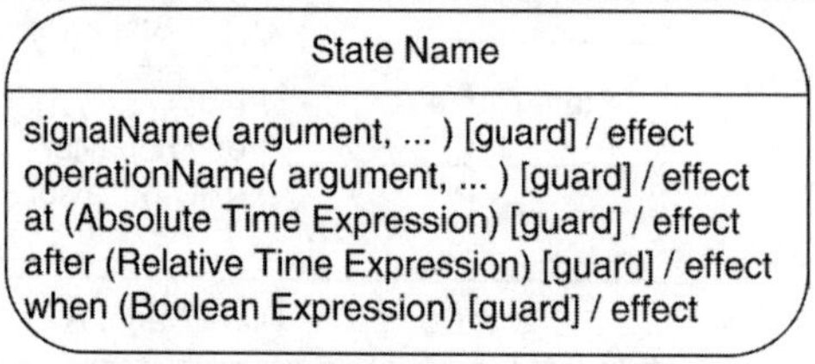	8.5.1, 8.5.2.1, 8.5.2.2, 8.5.2.3, 8.5.2.4
Initial Pseudostate		8.6
Junction Pseudostate	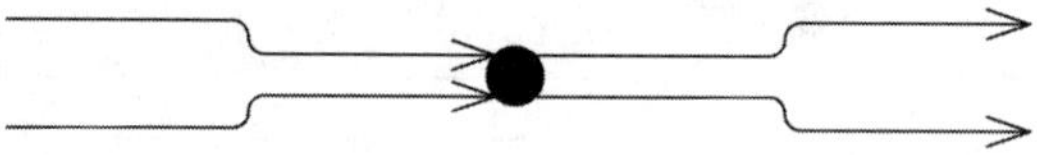	8.6

Table A.7 *Graphical Notations for Parametric Diagrams*

Element Kind	**Notation**	**Section**
Constraint Property		9.5
Constraint Parameter	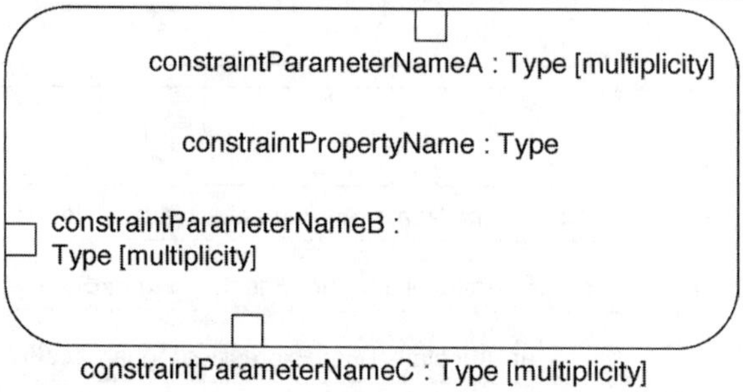	9.6
Value Property	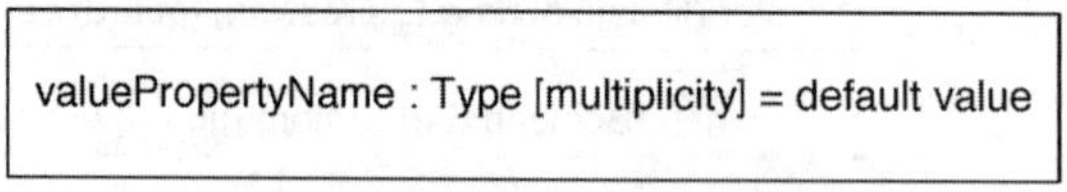 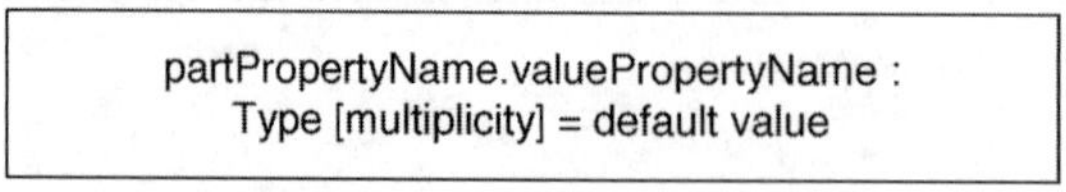	9.7

Table A.7 *(continued)*

Element Kind	Notation	Section
Binding Connector	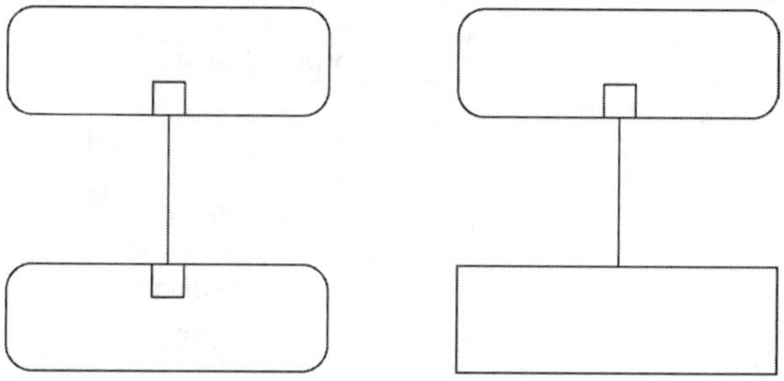	9.8

Table A.8 *Graphical Notations for Package Diagrams*

Element Kind	Notation	Section
Package	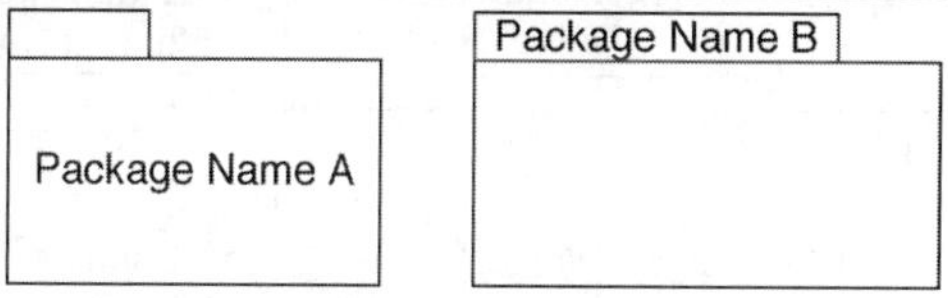	10.4
Model	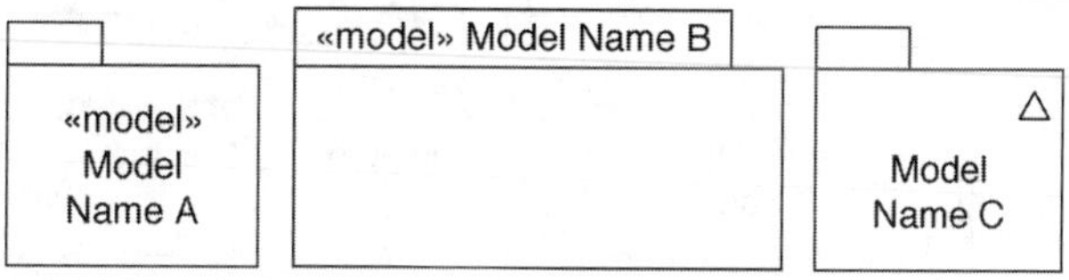	10.7.1
Model Library	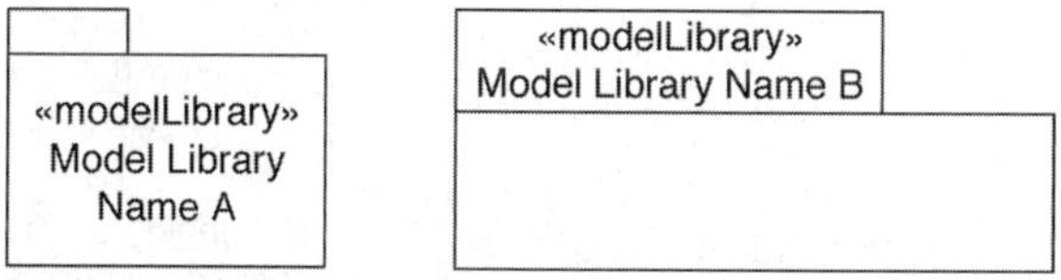	10.7.2
Profile	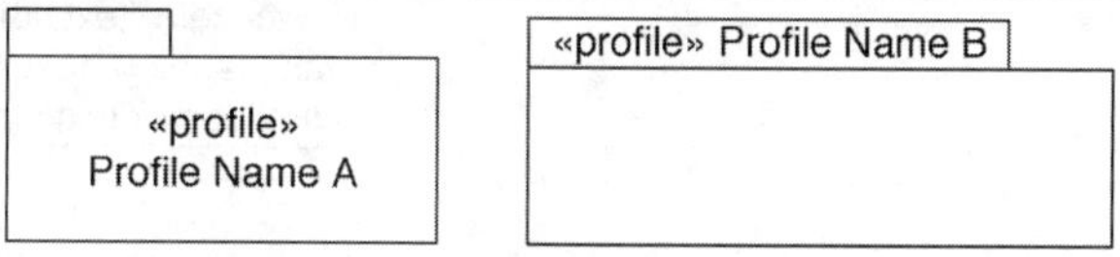	10.7.3
View	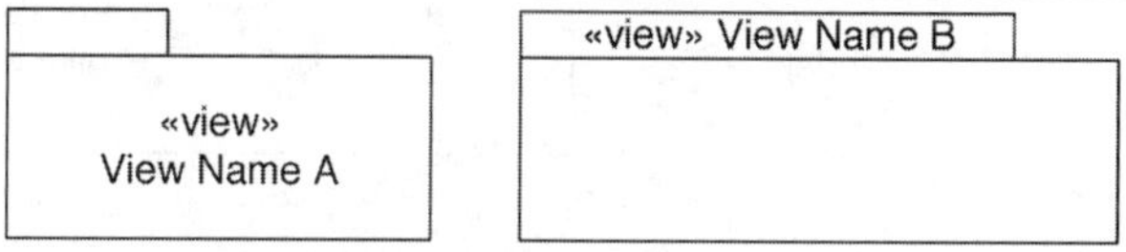	10.7.4

(continues)

Table A.8 *(continued)*

Element Kind	Notation	Section
Viewpoint	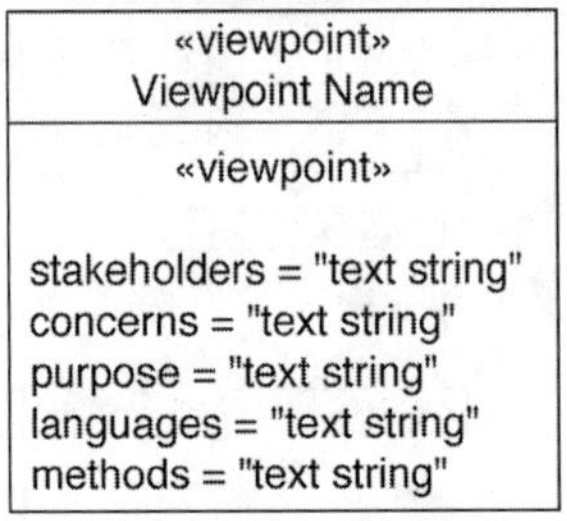	10.7.4
Namespace Containment	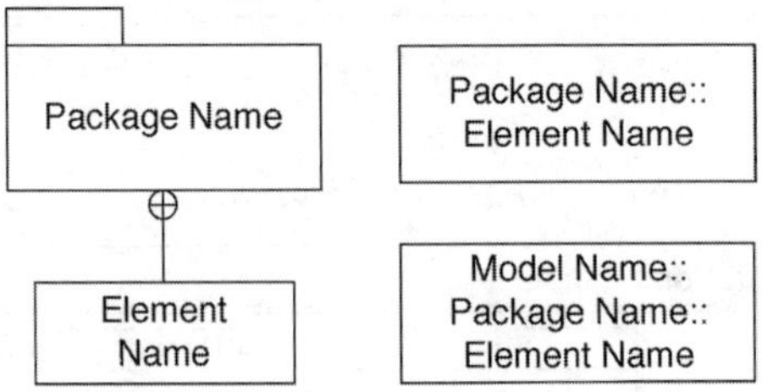	10.4
Dependency	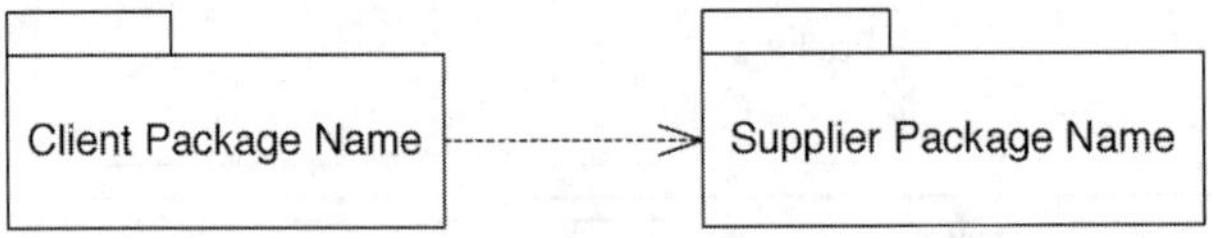	10.5
Package Import	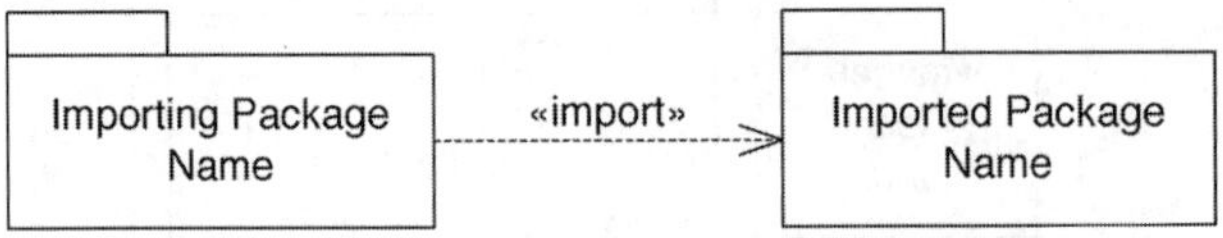	10.6
Conform	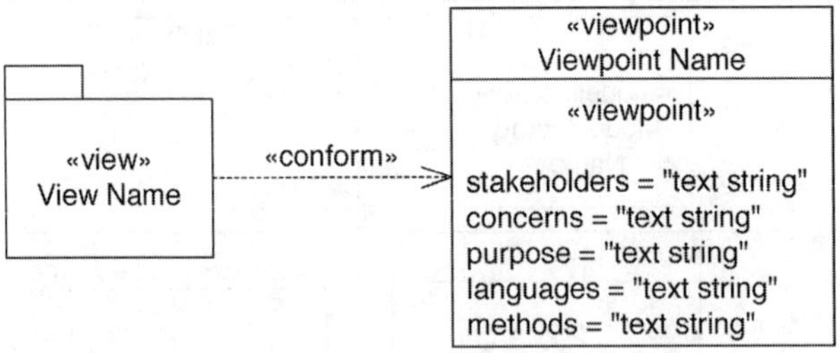	10.7.4
Profile Application	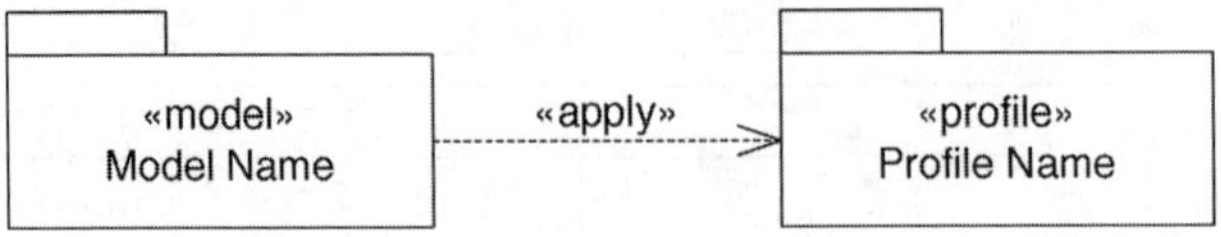	10.7.3

Table A.9 *Graphical Notations for Requirements Diagrams*

Element Kind	Notation	Section
Requirement	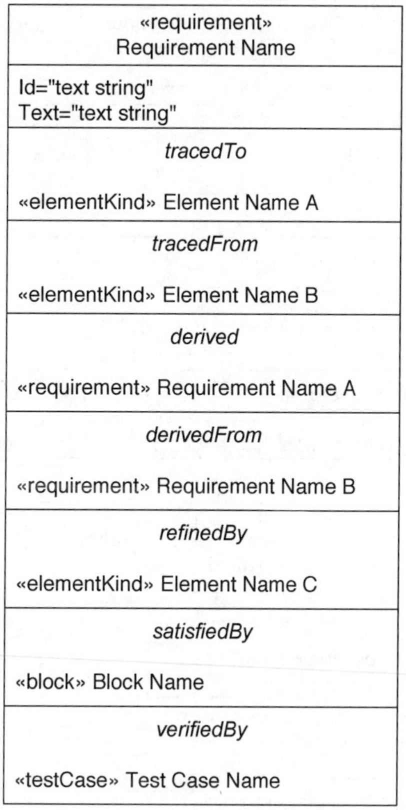	11.4
Rationale	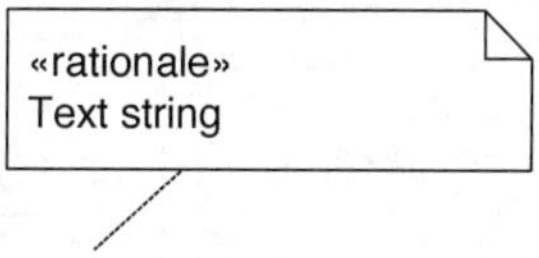	11.7
Namespace Containment	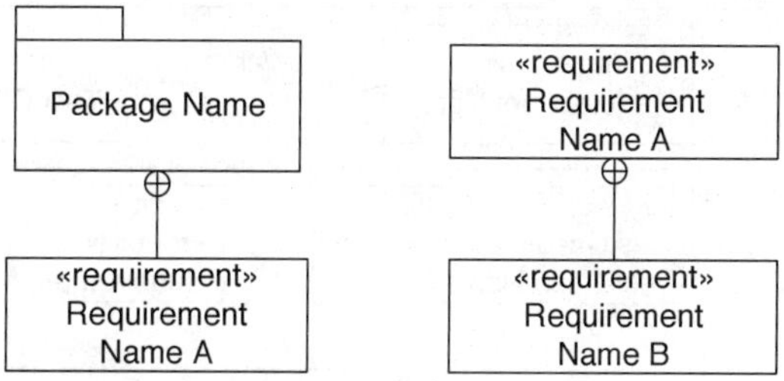	11.5

(*continues*)

Table A.9 *(continued)*

Element Kind	Notation	Section
Trace (Direct Notation)	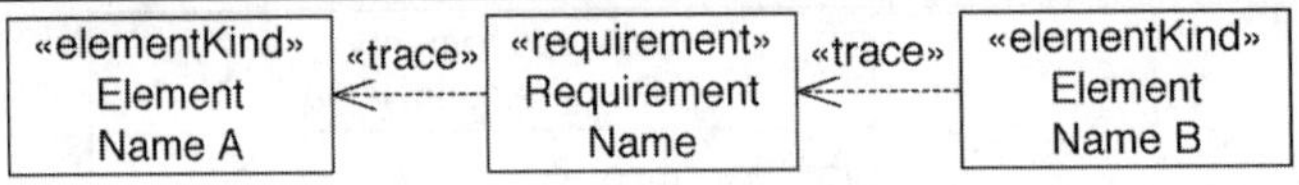	11.5
Trace (Callout Notation)	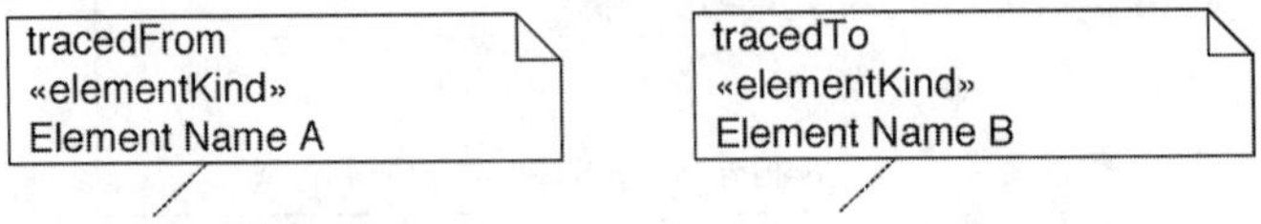	11.6
Derive Requirement (Direct Notation)		11.5
Derive Requirement (Callout Notation)	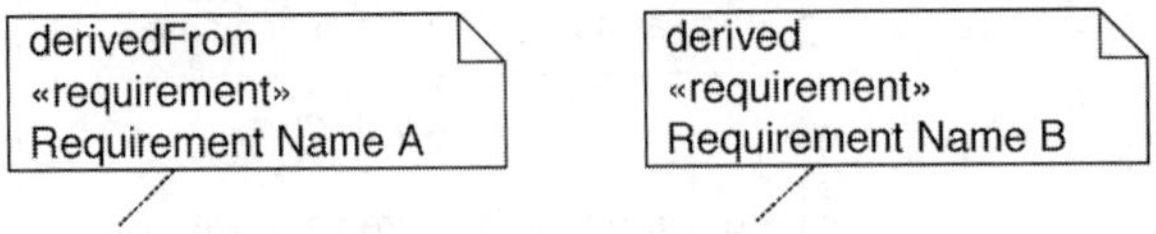	11.6
Refine (Direct Notation)	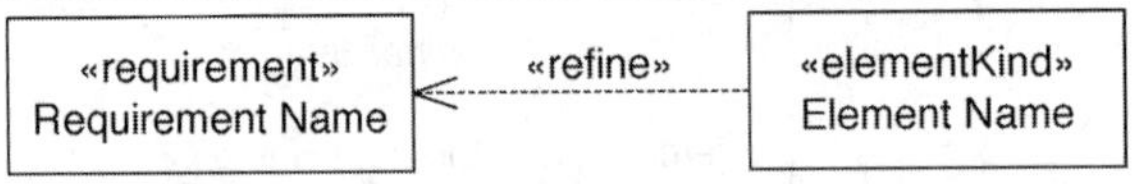	11.5
Refine (Callout Notation)	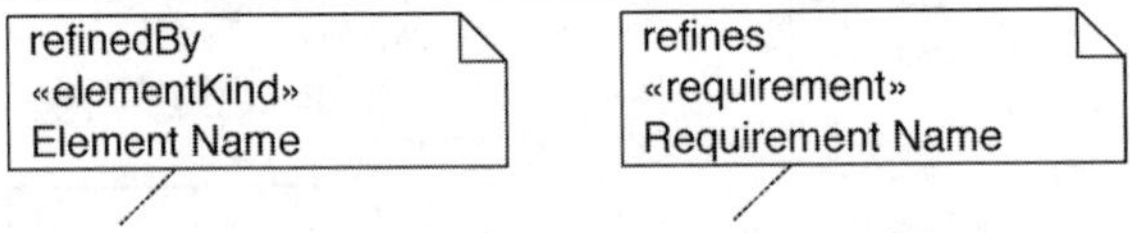	11.6
Satisfy (Direct Notation)	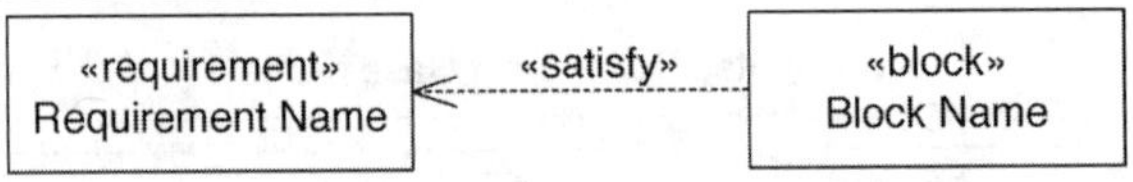	11.5
Satisfy (Callout Notation)	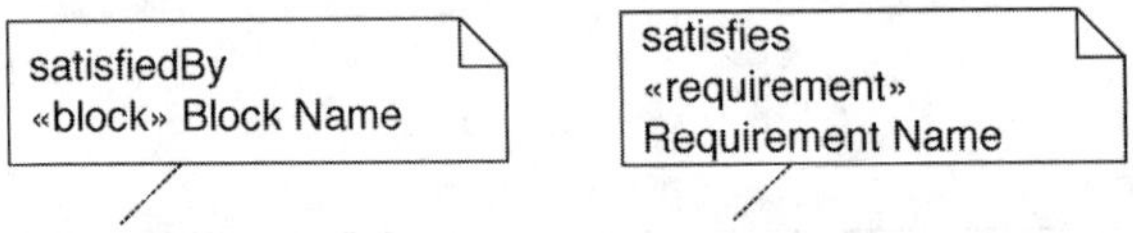	11.6
Verify (Direct Notation)	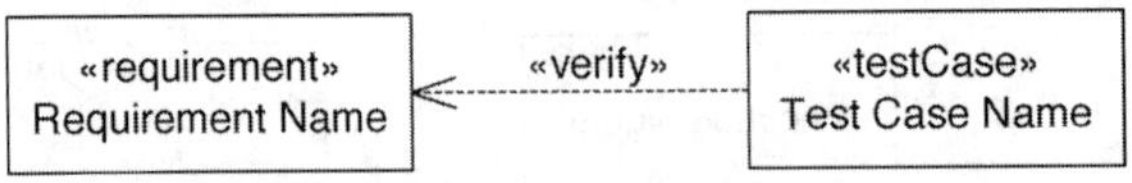	11.5
Verify (Callout Notation)	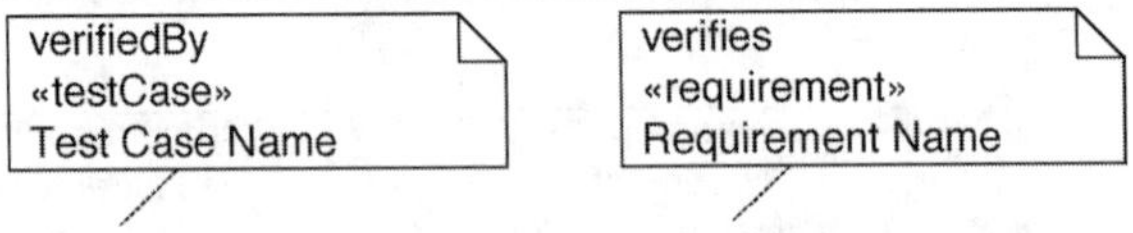	11.6

Table A.10 *Graphical Notations for Allocations*

Element Kind	Notation	Section
Allocation (Direct Notation)	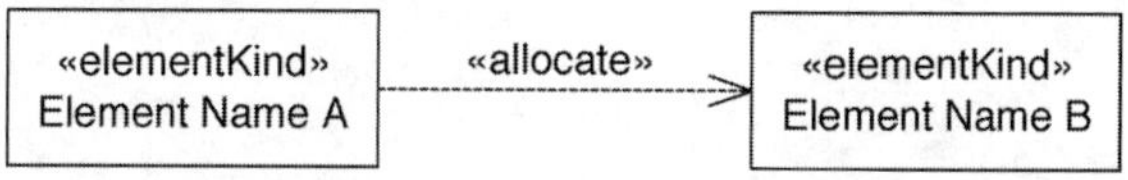	12.3
Allocation (Compartment Notation)	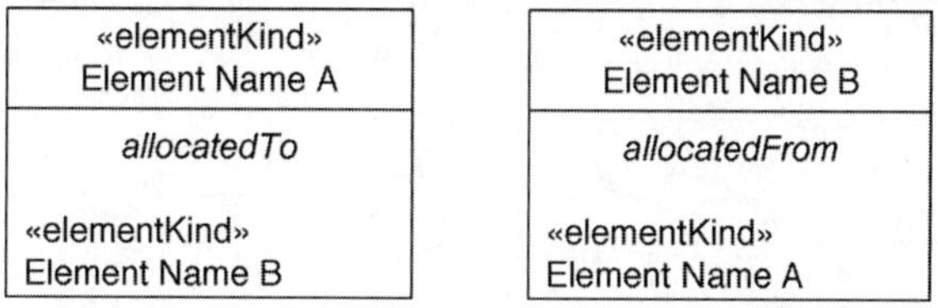	12.4
Allocation (Callout Notation)	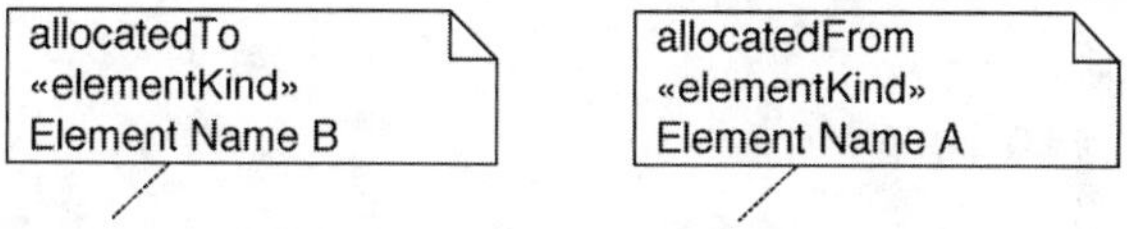	12.4
Allocation Activity Partition	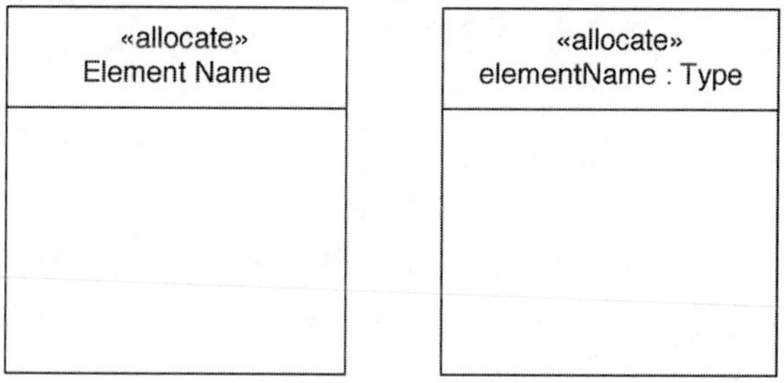	12.4
Rationale	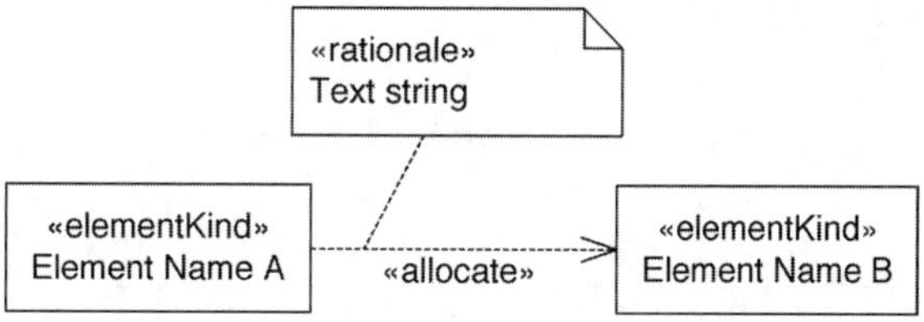	12.5

Appendix B

Changes between SysML Versions

SysML is a living language; it evolves continually to resolve errors and ambiguities in the specification and to better meet the needs of the systems modeling community. All practitioners have the right and the ability to submit issues about the SysML specification to the OMG. If you discover any errors or ambiguities in the specification or if you feel that the current version of SysML lacks a modeling feature you need for your work, you can write up the issue and submit it by e-mail to issues@omg.org.

A SysML Revision Task Force (RTF) convenes on an eighteen- to twenty-four-month cycle to review, discuss, and vote on a final disposition of the issues submitted during the preceding cycle. The result is a revision to the SysML specification incorporating the approved changes. The OMG then posts the finalized revision on its website, granting free access to all members of the modeling community.

The body of this book focuses on SysML v1.2, which the OMG published in June 2010. At the time of this writing, the current version of SysML is v1.3, which the OMG published in June 2012. Some modeling tools have already implemented the new language features introduced in SysML v1.3; others lag behind. All of them, however, continue to support SysML v1.2 features, including the small subset of features that became deprecated in v1.3. (And modeling practitioners who created models prior to v1.3 continue to use those features in their models.) For these reasons, I focus on SysML v1.2 in the body of this book; I discuss the differences between v1.2 and v1.3 in this appendix.

The biggest changes introduced in v1.3 (and the ones most likely to impact your daily work) are in the "Ports and Flows" clause of the SysML specification. In Chapter 3, "Block Definition Diagrams," and Chapter 4, "Internal Block Diagrams," I discuss standard ports, flow ports, and flow specifications. These language features are deprecated in SysML v1.3 and are retained solely for backward compatibility.

SysML v1.3 no longer distinguishes between *standard ports* (as a means to specify the services that are provided and required) and *flow ports* (as a means to specify flow properties). These concepts are replaced by the unified concept of a *port*, which consolidates the features of standard ports and flow ports into one; any port that a block owns can now specify the flow properties across that port as well as the services provided and required at that port.

As it did in SysML v1.2, a port still represents a distinct interaction point at the boundary of its owning block. The difference now is that a port is always typed by some other block that you've defined somewhere in the model hierarchy. The port therefore represents one or more instances of that other block—instances that exist at the boundary of the first block (the block that owns the port).

The BDD in Figure B.1 displays a block named *Solar Panel*. This block serves as the type of the *solarArray* port, which is owned by the *DellSat-77 Satellite* block. The *solarArray* port has a multiplicity of 2. This model conveys that each instance of *DellSat-77 Satellite* will own two instances of *Solar Panel*—and those two instances will exist at the boundary between the satellite system and its environment.

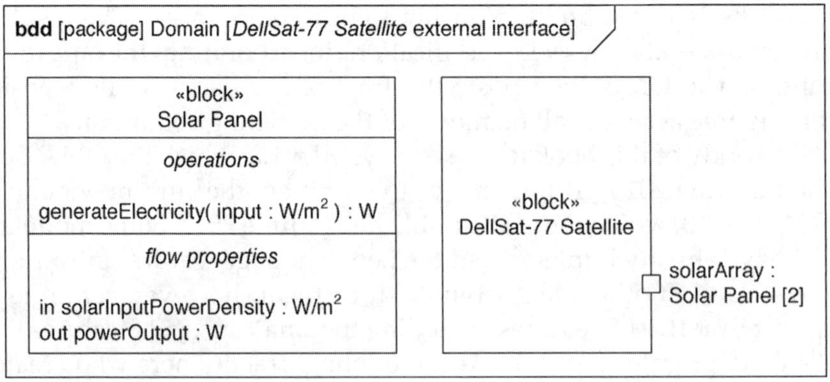

Figure B.1 *Ports in SysML v1.3*

In SysML v1.2, flow specifications (which type nonatomic flow ports) own flow properties. In SysML v1.3, flow specifications (as well as flow ports) are no longer supported; blocks are now augmented with the ability to own flow properties directly. And you can optionally display them in the flow properties compartment of a block.

In Figure B.1, the *Solar Panel* block displays three features that it owns: one operation and two flow properties. Because the *Solar Panel* block types the *solarArray* port, those three features are accessible to all structures that interface with the satellite via its solar array; those structures can invoke the *generateElectricity* operation, and they can exchange items with the satellite that are consistent (in direction and type) with the flow properties.

The IBD in Figure B.2 displays an *ssc* part property (of type *Solar Simulation Chamber*) and a *ds77* part property (of type *DellSat-77 Satellite*). A connector joins the *radiationSource* port (on *ssc*) to the *solarArray* port (on *ds77*). An item flow exists on this connector, which conveys that values of type W/m^2 can flow from *radiationSource* to *solarArray* during system operation. And this item flow is consistent in direction and type with the *solarInputPowerDensity* flow property of the *Solar Panel* block.

Figure B.2 also shows a part property, *db* (of type *Distribution Bus*), nested inside *ds77*. More precisely, the use of dot notation conveys that *ds77* owns a part property, *eps*, which in turn owns the part property *db*. A connector joins the *solarArray* port (on the boundary of the satellite) to the *db* part property (internal to the satellite). This connector has an item flow, *W*, which conveys that the output of the solar array is routed internally to the satellite's distribution bus. And this item flow

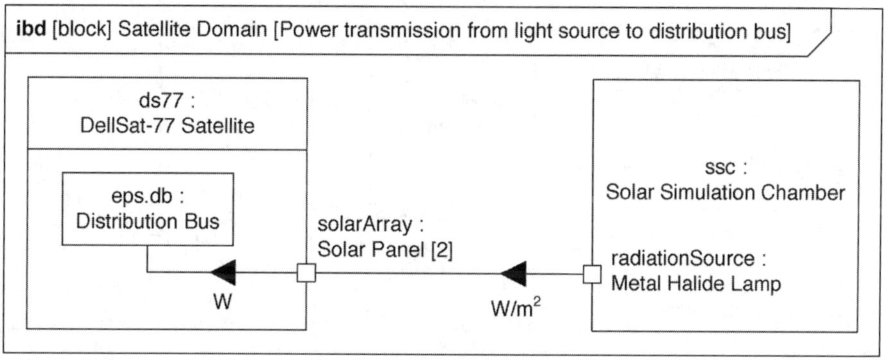

Figure B.2 *Item flows between ports and properties*

is consistent in direction and type with the *powerOutput* flow property of the *Solar Panel* block.

Another change that SysML v1.3 introduces is a port nesting capability; you can display a port on the boundary of another port. The BDD in Figure B.3 shows that the *solarArray* port has a nested port, *cells*. This conveys that the *cells* port is owned by the *Solar Panel* block, which types the *solarArray* port. (I display the *Solar Panel* block with its *cells* port on the same diagram for instructional purposes; it's redundant in this case.)

I recommend displaying nested ports sparingly. The port labels take up a lot of real estate on a diagram. Any value you gain from showing nested ports will be lost if your diagram becomes unreadable. Nested ports can be useful on IBDs when a single connector attached to a single port doesn't convey a sufficient degree of detail for your target audience. Showing nested ports allows you to display multiple connectors and specific attachment points for each one.

Thus far in this appendix, I've covered the key ideas about this new (unified) port concept in SysML v1.3. As mentioned earlier, SysML simply refers to it as a "port" (no qualifier required). However, I refer to it as a **non-stereotyped** port. That qualifier allows me to discuss the next topic more clearly.

A non-stereotyped port will meet your modeling needs in most cases where you feel a port is even necessary. However, you should know that SysML v1.3 introduces two new specialized kinds of ports—full ports and proxy ports—that you can use if you feel that a non-stereotyped port doesn't provide a sufficient degree of fidelity.

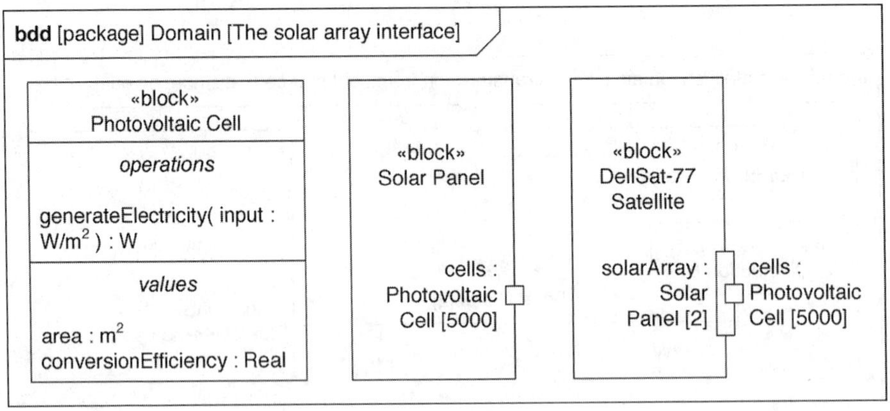

Figure B.3 *A port with a nested port*

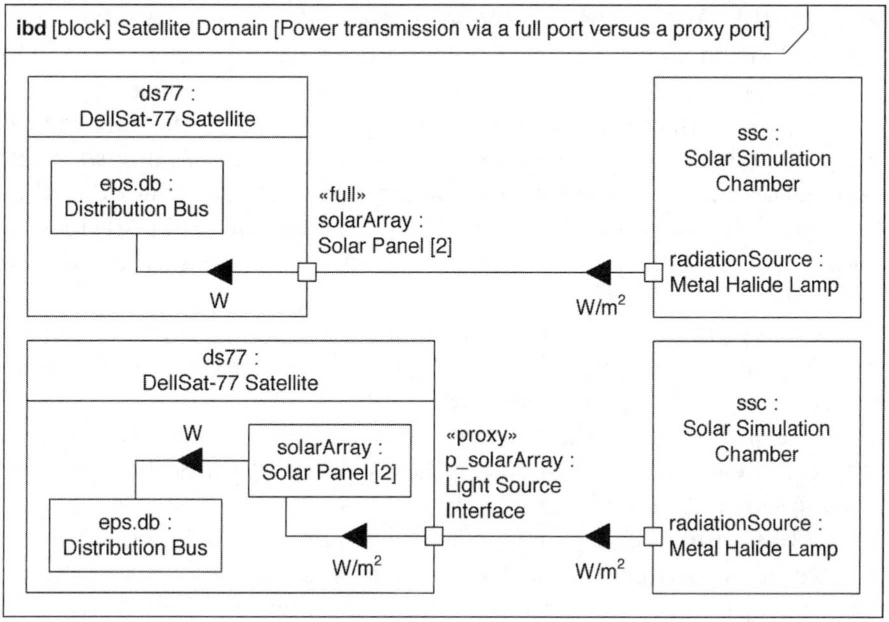

Figure B.4 *Full ports versus proxy ports*

The IBD in Figure B.4 shows a full port at the top of the diagram and a proxy port at the bottom. A full port has the stereotype «full» at the beginning of the name string; a proxy port has the stereotype «proxy» at the beginning of the name string. Otherwise, the notations for these kinds of ports are the same as for a non-stereotyped port.

A **full port** represents a part property of the owning block—a part property that just happens to exist at the boundary of that block. Like other part properties, a full port is typed by a block. And like all part properties, a full port can own and perform behaviors and can have an internal structure (i.e., it can have its own nested part properties).

Applying the «full» stereotype to the *solarArray* port in Figure B.4 conveys that the *DellSat-77 Satellite* block owns a part property named *solarArray* (of type *Solar Panel* with a multiplicity of 2). It is in every sense a part property, just like the nested *db* part property. Modeling *solarArray* as a full port (instead of as a nested part property) simply conveys the additional piece of information that it exists at the satellite's boundary (i.e., it partially defines the satellite's external interface).

As a part property, the *solarArray* full port can own and perform behaviors (for example, the *generateElectricity* operation in the *Solar*

Panel block). This means that the *solarArray* full port can receive items, act on them in some way that potentially transforms them, and then output other items (possibly of types that differ from those of the inputs it received). Figure B.4 shows that the *solarArray* full port can receive instances of the value type W/m^2 from the external environment and can output instances of the value type *W* to the satellite's distribution bus. This conveys definitively that the *solarArray* full port does, in fact, perform a behavior that transforms its input into a different type of output.

A proxy port is fundamentally different. It does not represent a part property of the owning block. In fact, it doesn't represent anything physical. Instead, a **proxy port** represents its owning block's external interface. More precisely, it represents the subset of behavioral and structural features of the owning block (or its part properties) that are accessible to external blocks.

Unlike a full port, a proxy port neither performs behaviors nor has an internal structure (i.e., nested part properties). It is what it sounds like: It's a proxy for other things—either the owning block itself or its internal parts. The best comparison I can offer is that a proxy port is like a portal at the boundary of a block through which that block can exchange a subset of services and flow properties with external blocks.

Applying the «proxy» stereotype to the *p_solarArray* port in Figure B.4 conveys that this port does not represent a part that can perform behaviors or send and receive flow properties. Rather, the *p_solarArray* proxy port simply defines an interaction point between an external object (a solar simulation chamber) and one of the satellite's internal parts (the solar array). Instances of the value type W/m^2 that arrive at the proxy port are simply routed to the *solarArray* part property. The proxy port itself cannot act on those instances in any way. The behaviors that are invoked at that port are actually performed by the *solarArray* part property.

A proxy port is typed by a specialized kind of block called an **interface block** (which is also new in SysML v1.3). The notation for an interface block is the same as the one for a block, but with the «interfaceBlock» stereotype preceding the name of the element. You would create an interface block in your model to specify the subset of behavioral and structural features of another block that are accessible at a particular proxy port.

The BDD in Figure B.5 displays an interface block named *Light Source Interface*. This interface block contains a subset of the features shown in the *Solar Panel* block: one operation and one flow property.

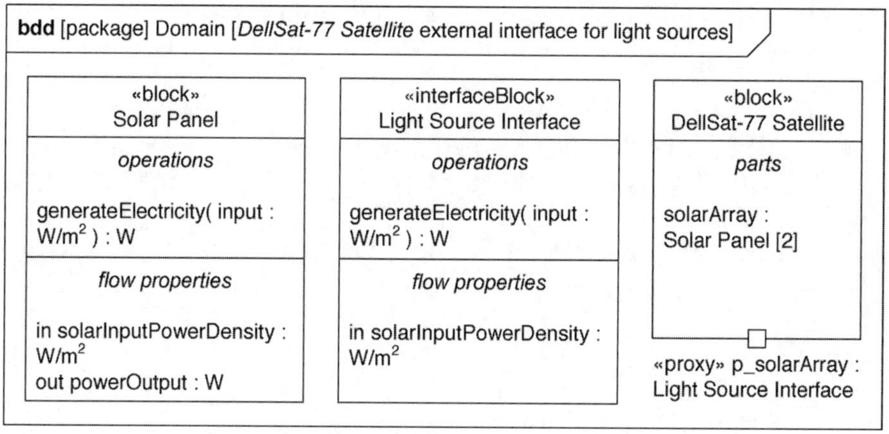

Figure B.5 *An interface block and a proxy port*

Light Source Interface types the *p_solarArray* proxy port on the *DellSat-77 Satellite* block. This view of the model (coupled with the view shown in Figure B.4) conveys that the satellite owns an internal part (*solarArray*) and offers external access to two of its features via the *p_solarArray* proxy port.

The choice to use a non-stereotyped port, a full port, or a proxy port is purely a methodological decision. My preference is to use non-stereotyped ports unless there's a compelling reason to use one of the specialized kinds. Your team should have a modeling standards and conventions document to guide all members of the team in modeling consistently.

Bibliography

Baker, Paul, et al. *Model-Driven Testing: Using the UML Testing Profile*. Berlin: Springer, 2010.

Brooks, Jr., Frederick. *The Design of Design: Essays from a Computer Scientist*. Boston: Addison-Wesley, 2010.

Cockburn, Alistair. *Writing Effective Use Cases*. Boston: Addison-Wesley, 2001.

Dori, Dov. *Object-Process Methodology: A Holistic Systems Paradigm*. Berlin: Springer-Verlag, 2002.

Estefan, Jeff. *Survey of Model-Based Systems Engineering (MBSE) Methodologies*, Rev. B. Seattle: INCOSE, 2008. Available at http://www.incose.org/ProductsPubs/pdf/techdata/MTTC/MBSE_Methodology_Survey_2008-0610_RevB-JAE2.pdf.

Fortescue, Peter, et al. *Spacecraft Systems Engineering, Third Edition*. Chichester: John Wiley & Sons, 2003.

Fowler, Martin. *UML Distilled: A Brief Guide to the Standard Object Modeling Language*. Third Edition. Boston: Addison-Wesley, 2004.

Friedenthal, Sanford, et al. *A Practical Guide to SysML, Second Edition: The Systems Modeling Language*. Boston: MK/OMG Press, 2011.

Haskins, Cecilia, ed. *INCOSE Systems Engineering Handbook: A Guide for System Life Cycle Processes and Activities*, v3.2.2. International Council on Systems Engineering (INCOSE), October 2011.

Hoffmann, Hans-Peter. "Harmony-SE/SysML Deskbook: Model-Based Systems Engineering with Rhapsody," Rev. 1.51, Telelogic/I-Logix white paper. Telelogic AB, May 2006.

Hosken, Robert W., and James R. Wertz. "Microcosm Autonomous Navigation System On-Orbit Operation." Available at http://www.microcosminc.com/analysis/manspaper.pdf.

International Council on Systems Engineering. *Systems Engineering Vision 2020*, v2.03. September 2007. Available at http://www.incose.org/ProductsPubs/pdf/SEVision2020_20071003_v2_03.pdf.

Maier, Mark W., and Eberhardt Rechtin. *The Art of Systems Architecting, Third Edition*. Boca Raton, FL: CRC Press, 2009.

Object Management Group. *OMG Systems Modeling Language (OMG SysML)*, v1.2. June 2010. Available at http://www.omg.org/spec/SysML/1.2/.

Object Management Group. *OMG Systems Modeling Language (OMG SysML)*, v1.3. June 2012. Available at http://www.omg.org/spec/SysML/1.3/.

Object Management Group. *OMG Unified Modeling Language (OMG UML)*, v2.4.1. August 2011. Available at http://www.omg.org/spec/UML/2.4.1/.

Piscitelli, Roberta, and Gianluca Furano. "Definition of a Cryptographic Block for Satellite Telemetry." Available at http://staff.science.uva.nl/~roberta/publications_files/guide.pdf.

Rumbaugh, James, Ivar Jacobson, and Grady Booch. *The Unified Modeling Language Reference Manual, Second Edition*. Boston: Addison-Wesley, 2005.

Seidewitz, Ed. "What do models mean?" March 2003. Available at http://www.semanticcore.org/Docs/WhatDoModelMean.pdf.

Sellers, Jerry Jon. *Understanding Space: An Introduction to Astronautics, Third Edition*. New York: McGraw-Hill, 2005.

Simsion, Graeme C., and Graham C. Witt. *Data Modeling Essentials, Third Edition*. San Francisco: Elsevier, 2005.

Spinellis, Diomidis, and Georgios Gousios. *Beautiful Architecture: Leading Thinkers Reveal the Hidden Beauty in Software Design*. Sebastopol, CA: O'Reilly Media, 2009.

Weilkiens, Tim. *Systems Engineering with SysML/UML: Modeling, Analysis, Design*. Boston: MK/OMG Press, 2008.

Wertz, James R., and Wiley J. Larson. *Space Mission Analysis and Design*. Third Edition. Hawthorne, CA: Microcosm Press, 1999.

Yourdon, Edward. *Modern Structured Analysis*. Englewood Cliffs, NJ: Prentice Hall, 1989.

Index

D

J

L

M

N

T

U